D0397058

QUANTUM FUZZ

QUANTUM FUZZ

THE STRANGE TRUE MAKEUP OF EVERYTHING AROUND US

MICHAEL S. WALKER

59 John Glenn Drive
Amherst, New York 14228

Published 2017 by Prometheus Books

Cover image from NASA/CXC/M. Weiss
Cover design by Jacqueline Nasso Cooke
Cover design © Prometheus Books
Interior cartoons illustrated by Sidney Harris, ScienceCartoonsPlus.com.

Inquiries should be addressed to
Prometheus Books
59 John Glenn Drive
Amherst, New York 14228
VOICE: 716–691–0133
FAX: 716–691–0137
WWW.PROMETHEUSBOOKS.COM

21 20 19 18 17 5 4 3 2 1

Library of Congress Cataloging-in-Publication Data

Names: Walker, Michael S., 1939- author.
Title: Quantum fuzz : the strange true makeup of everything around us / by Michael S. Walker.
Description: Amherst, New York : Prometheus Books, [2017] | Includes bibliographical references and index.
Identifiers: LCCN 2016031862 (print) | LCCN 2016035281 (ebook) | ISBN 9781633882393 (hardcover) | ISBN 163388239X (hardcover) | ISBN 9781633882409 (ebook)
Subjects: LCSH: Quantum theory.
Classification: LCC QC174.12 .W347 2017 (print) | LCC QC174.12 (ebook) | DDC 530.12—dc23
LC record available at https://lccn.loc.gov/2016031862

Printed in the United States of America

For my family, passed and present

CONTENTS

FOREWORD

We live in a special time in history. The knowledge of the world is expanding at an exponential rate. Perhaps more important, to quote one of my favorite scientists of all time, Carl Sagan: "We live in a society exquisitely dependent on science and technology, in which hardly anyone knows anything about science and technology."[1] This is a real problem and it needs to be fixed.

And, beyond that, there is an excitement to science that most people miss out on. Part of the problem is that mathematics truly is the language of science, and the mathematics involved has become so complex that only those with a high degree of training can appreciate science in its native language. But now and then a "translator" comes along who can convey the meaning, the beauty, and the excitement to the rest of us.

What Mike Walker has done in this book places him on the cutting edge as such a translator. Though he is a nonacademic, through this popular book he has conveyed a clear understanding of the physical world, while circumventing the math. As he demonstrates, you don't HAVE to go to a classroom to understand the excitement. Then again, going to the classroom these days might not even GET you the excitement because you'll get mired in the math, or they just don't teach the fun stuff discussed here—even if you are majoring in physics. While some of the details will not stand the test of time (which is true in any cutting-edge field), I feel comfortable recommending this book also to all students, as a supplement to the normal text so that they may more fully understand what is going on beyond the math. *Quantum Fuzz* is not designed to replace course textbooks, but it sure is a lot more fun, readable, and interesting.

At the core of this book is quantum mechanics. While it surely goes beyond the ideas, and well into the applications, Walker brings to life one of the most strange, fascinating, and beautiful descriptions of our physical world. He takes great pains to describe that part of quantum mechanics particular to the atom, and he uses it as a basis for explaining chemistry and everything around us. It is this part of physics that is "translated" for us through the nonmathematical and largely visual presentations of *Quantum Fuzz*. This book further "comes alive" through sections now and then relating to the history of the development of physics and chemistry, and to the lives of the many prominent scientists most responsible for moving this history along.

I like the descriptions of how troubling early scientists found quantum mechanics to be, despite its incredible successes at predicting the outcomes of experiments. I love the stories of some of the great debates between science's giants. For example, Albert Einstein (who hated quantum mechanics and its counterintuitive conclusions but ultimately won a Nobel Prize for his work in the area), is reputed to have said "God doesn't play dice with the Universe," and Niels Bohr (another Nobel laureate) responded with the not-quite-as-famous-but-equally-important retort, "Don't tell God what to do." Bohr was also to have said, "Anybody who is not shocked by this subject has failed to understand it." And yet here it is, more than eighty years later, and quantum mechanics is arguably the most successful, debated, and tested theory in history.

This is not the first book that interweaves science and history. Indeed, most nonacademic books on this subject do so as a way of softening the conceptual challenge. But Walker also explains chemistry as it follows from the physics, and how the understanding of quantum mechanics has made possible much of modern invention and technology. He does this from the vantage point of a lifetime of work in industrial laboratories (and even includes brief descriptions of some of the projects that he himself has worked on). And he goes beyond what we see here on Earth to show how we use our knowledge and understanding of quantum mechanics to interpret what we see of the cosmos, to understand how the universe began with a big bang and came to be the way it is today.

Ultimately, human beings and the things here on Earth are all made of atoms. Yet most people know nothing of their diffuse, fascinating symmetries, and how these forms determine much of the properties of our universe. This book is an opportunity to come on board and sail to new lands of understanding.

I believe that you will enjoy this book as I did. And I am sure that you will agree that scientific fact is, indeed, much stranger than fiction.

David Toback
Thaman Professor for Undergraduate Teaching Excellence
Professor of Physics and Astronomy
Mitchell Institute for Fundamental Physics and Astronomy
Texas A&M University
September 2016

AUTHOR'S NOTE

Professor Toback teaches a course in cosmology for nonscience majors and in 2013 published the very readable *Big Bang, Black Holes, No Math.*

PREFACE

- I look outside at the snow, frozen water, H_2O molecules. But for their quantum makeup, they wouldn't exist (and I wouldn't exist).
- I answer my cell phone. LCD display, wires, semiconductor chips. But for their quantum makeup, none of these would exist.

We live in a beautiful, fascinating, quantum world. We ourselves are quantum beings. All life and matter are quantum, and our technologies advance more and more based on our understanding of quantum principles. Yet most of us have only the vaguest sense of any of this.

In 1900, the German physicist Max Planck found that light is radiated from hot objects in chunks of energy, which he called "quanta." This proved to be only the tip of the iceberg in a journey of discovery that has led to terms such as *quantum revolution*, *quantum theory*, *quantum mechanics*, and *quantum world*.

I have written this book to provide for the thoughtful general reader a readily understandable view into this quantum world. I present this view in the context of the rich history of discovery and conflict in science and human events experienced during the last one hundred and twenty years. To accomplish this task I have unabashedly borrowed and simplified from the best of what has been written or otherwise presented on this subject, in each case studiously avoiding any but the simplest of math that may have been involved.

Realize that, were this not a quantum world, atoms would be able to overlap each other, so the volume of any substance composed of atoms would be reduced perhaps a million billion times, but its weight would be the same as it is now. We would be tiny, only as tall as the thickness of a human hair, but still have our same weight. And we would have a different chemistry: elements would have entirely different appearances and properties. Molecules, if they existed, would be much different: no water, no air, perhaps only solids—atoms perhaps bound together only by gravitational attraction. Maybe no fire. Living things? In what form? Thought? Consciousness?

The clumping together of substances to form stars and galaxies would have transpired differently and would have taken much longer were it not for the unevenness in the early universe, possibly explained by quantum fluctuations. We probably wouldn't have our sun. No sunlight. No hospitable Earth.

Yet, we are as we are, galaxies do exist, and we have what surrounds us because of the quantum nature of the universe and the atom. Although we don't see it in our everyday lives, the inner workings of this world are very strange, strange to the point that the mind boggles.

"Quantum theory," in its more mathematical formalism as "quantum mechanics," explains not only this strangeness, but, together with Einstein's relativity, every observed aspect of the world around us, including the failings of classical Newtonian physics. "Not a single one of the theory's predictions has ever been shown wrong,"[1] and it has opened new avenues for invention. Even five years ago it could be said that fully "one third of our economy depends on products based on it."[2]

In regard to strangeness, first realize that events in our quantum universe are no longer absolutely determined and predictable from past events but instead are based on probabilities. There is "entanglement," non-locality, what Einstein called "spooky action at a distance."[3] A measurement here on one object can instantaneously, faster than the speed of light, determine the outcome of a measurement made on another object far away, with nothing carrying the information in between. Our notion of cause and effect has to be changed. (Interestingly, the solution to all of this strangeness may be stranger still: that we live in just one of many parallel worlds.)

So how is it that engineers, with all of this probability and strangeness, have been able to project the motion of the planets, send rockets to the moon and beyond, and build machines that run with seeming precision? Well, the familiar classical ideas that they have used (and will continue to use) to describe the macro world can be viewed as a kind of physical shorthand. These ideas provide a practical, easier, and quicker way of very closely approximating for large objects (where even a tiny grain of sand is a very large object) the more detailed and precisely correct (but more cumbersome and often unnecessary) probabilistic workings of quantum physics.

Realize, however, that the modern tools that these engineers use for design and some of the components in their machines are actually products of the quantum view. These include the laser, superconductors, and all of the solid-state semiconductor electronic devices of the modern day.

In the chapters that follow, I explain how we have come to understand and use the quantum properties of our world to our advantage. In a chronological narrative starting in Part One, I first describe the "quantum revolution" that led to this understanding, and the controversy involved. I relate key experimental and theoretical results, including the discovery of the quantum, the probabilistic nature of physics, and the now accepted scientific view of the simplest of atoms, showing for example how these concepts led to the invention of the laser with all of its applications.

In Part Two, I briefly summarize the further development of the theory and describe its broader implications: entanglement, that events can't be both "objectively real" and "local," and what we mean by these terms. Here, I also describe just a few potential new applications: superpowerful quantum computers that will operate for some applications at lightning speeds compared to conventional computers, quantum cryptography, and, in a flight of fancy, the prospect of entangled teleportation. (It's not what you think it is.)

In Part Three I provide a glimpse of the quantum's broad influence, from the tiniest of particles after the big bang to the formation of the stars and the galaxies. We consider black holes, supernovae, the Higgs boson, and the fundamental particles of our universe. Here I indicate the seeming incompatibility of quantum mechanics and general relativity at the intense energies following the big bang, with reference to sources on string theory and loop quantum gravity as possible solutions.

In Part Four, I make the connection from quantum theory to the practical world and what we see around us. I relate how the form and physics of many-electron atoms follows from the quantum configuration of the hydrogen atom, to explain chemistry and the nature of all substances. (For me this is the most important product of quantum mechanics: the provision of an understanding that is the practical engine of invention.)

I paint a picture of what happens inside the atom in a way that may even be enlightening to students and graduates in physics and chemistry. I conclude with short chapters introducing a bit about bonding and materials science, as background to the quantum wonders to be described in Part Five.

In this final part of the book I describe many of the inventions in materials and devices that have resulted from, or are explained by, our knowledge of quantum mechanics. I include superconductors and superconducting devices, fusion power, and the solid-state electronic quantum devices of the modern day. I also indicate developments under way to produce new superconductors, semiconductors, and other materials, including graphenes and nanotubes, with applications in medicine, electronics, and energy storage.

Throughout the book I provide glimpses into the lives and personalities of the often brilliant men and women who have moved this narrative forward. To distinguish these and other interesting bits of information from the main text, I have boxed the regions in which these digressions occur. Here and there I also provide extra explanation or background, sometimes out of chronological sequence. These passages are indented. Finally, for those who would like to dig deeper, I provide (1) a series of appendices related to specific chapters and (2) a list of books (and one lecture series on CDs) that I would recommend. I often cite these same books, each labeled for easy reference with a letter from A through Z. For those books beyond the first twenty-six, I use two identical letters: AA, BB, CC, and so on.

I welcome you now to *Quantum Fuzz.*

Michael S. Walker, PhD

Part One

DISCOVERY AND UNDERSTANDING (1900–1927)

Fig. 1.1. Attendees of the Fifth Solvay International Conference, on quantum mechanics, October 24–29, 1927. (Photograph by Benjamin Couprie, courtesy of International Solvay Institutes.)

Chapter 1

INTRODUCTION TO PARTS ONE AND TWO

Ours is a quantum world, but the discovery of that world took many decades of experimental and theoretical work. Starting in 1900, a radical new theory was developed that explained the chemistry of the elements, the periodic table, the sizes of atoms, why we are the size that we are, and various phenomena that had defied explanation using the conventional *classical* view of the world that existed until that time (including, for instance, Newton's laws of motion that describe the falling apple and the orbits of the planets).

The new conceptual ideas are referred to broadly as *quantum theory*, and the mathematical approaches that were developed to describe and integrate these ideas into a generally applicable method of calculation are known as *quantum mechanics*. Collectively this body of work has been called "the most successful set of ideas ever devised by human beings"[1] and "the most powerful physical theory that has ever been devised."[2]

Until 1925, quantum theory was an assemblage of ad hoc postulates, assumptions, and quasi-classical constructs that managed to explain experimental findings. But in the next several years a firm foundation for the overall effort was laid in place when three young scientists developed separate mathematical constructs that accurately described the one-electron hydrogen atom.

In the fall of 1927, some twenty-four of the top scientists from around the world, "the greatest gathering of physicists ever,"[3] met for nearly a week in Brussels for the fifth conference sponsored by the Belgian industrialist Ernest Solvay, this one devoted exclusively to examining these exciting, new, and continuing developments in quantum mechanics. This group and five other guests are shown in Figure 1.1. Seventeen of this group were by then or would later be recipients of the Nobel Prize in physics or chemistry. (Note: the Nobel Prize is awarded only to scientists who are alive at the time that the award is to be given. And the honor is usually bestowed many years after the work that merits it has been done. So many a worthy scientist has died before he might have received the prize.)

By the time of this Solvay conference, the physics community had divided into two camps with dramatically opposite views on the interpretation and implications of the theory: one camp was led by Albert Einstein (at the center of the first row in the figure) and the other was led by Niels Bohr (at the far right in the second row). The opposing views ran so deeply as to dispute the meaning of reality and physics itself. These views had recently been defined, but this was the first time that all of the major players on both sides would be assembled to present and discuss them. It was to be a clash of titans.

In Part One, I explain the lead-up to the meeting and how quantum theory and quantum mechanics were developed. I describe experiments, ideas, and the people involved. In Part Two, I describe the meeting and the controversy over the new theory's arrival, explore its mind-boggling implications, and tell of much later definitive experiments that would judge the debate.

Note that in providing the historical narrative for these Parts One and Two I draw heavily on the excellent book *Quantum—Einstein, Bohr, and the Great Debate about the Nature of Reality*, by Manjit Kumar.[4] And, at appropriate points throughout these same parts and also in Part Three, I indicate the awards of Nobel Prizes in Physics and recite the rationale that the Nobel Foundation gave for each award. My source for all of these citations is a section titled "Nobel Prize Winners," in *Physics: Decade by Decade*, by Alfred B. Bortz, a book of the Twentieth Century Science series.[5]

SCIENTIFIC NOTATION AND SCIENTIFIC SHORTHAND

Throughout this book, I avoid all but the simplest of mathematics. Instead, I use a "scientific shorthand" to describe a few simple physical relationships. And, because you will encounter here and there some very large and some very small numbers, I simplify by using a "scientific notation." A couple of examples of both the shorthand and the notation are provided and explained in the indented paragraphs that follow. I suggest that you take a couple of minutes now to look at these examples so that you will be familiar with both conveniences and can easily proceed in those instances where you may need them.

For example, I will refer to the speed of light as c, which we know from Einstein's formula $E = Mc^2$ for the energy equivalent of a mass of material. (This formula, or equation, is just a scientific shorthand.) And c is just a shorthand for the number and the units describing the speed of light. $c = 299{,}793{,}000$ meters per second, where 299,793,000 is the number and meters per second contains the units (meters and seconds) and is written in shorthand as *m/s*. (Note that a meter, the standard international [SI] unit of length, is just three inches longer than a yard. You may be familiar with the terms "yardstick" or "meter stick" to measure length.) The superscripted 2 after the c simply means that c is multiplied by itself, $c \times c$.

To concisely and sometimes approximately express large numbers such as c, we'll be using a scientific notation in which we: (a) round off to the first few digits expressed in decimal form (2.998, for our example), and then (b) multiply by the number ten raised to a superscript ("power") given by the total number of digits that would follow the decimal point. (This is the same as the number 10 multiplied by itself that many times.) So, in this notation, $c = 2.998 \times 10^8$ *m/s*. The number of digits shown may depend on the precision to which a quantity is known or the degree of precision needed in its use. Rounding off further, we would have the easy-to-remember $c = 3 \times 10^8$ *m/s*. (Here 10^8 represents ten multiplied by itself eight times.)

We will also encounter some very small numbers. For example, the mass of an electron is 0.00000000000000000000000000000091083 kilograms, where one kilogram is about 2.2 pounds. We write this as 9.1083×10^{-31} kg, where the minus sign in 10^{-31} indicates that 9.1083 is *divided* by 10^{31}, that is, divided by ten thirty-one times.

Chapter 2

PLANCK, EINSTEIN, BOHR—EXPERIMENTS AND EARLY IDEAS

The realization that we live in a quantum world came on slowly, with a discovery here and there that didn't seem to fit with the viewed-to-be-absolutely-correct and universally accepted classical physics of the time. We now examine these findings in the manner of a Sherlock Holmes, spotting clues and piecing them together to solve some great puzzle. As we move from clue to clue, I'll be providing boxed biographies and histories of the personalities involved in the search, and I'll also include bits of background on the physics, in indented paragraphs, so that the meaning of the clues can be understood. The first clue, what started it all, was examined by Max Planck.

THE QUANTUM (A RESULT OF THE FIRST CLUE)

Two chairs to the left of Einstein in the first row in Figure 1.1 is Marie Curie, and further next to her, second from the left, is Max Planck, who in 1900 started the quantum revolution with his explanation of the radiation of heat and light. At the time, the center of work on theoretical physics was in Germany, and in all of Germany there were only sixteen professors of theoretical physics. It was a small community. Most of the advancements in theoretical physics would come from men who were in their twenties, many of them in their early twenties. Planck was over forty.

Recognizing that advantages were to be gained through the development of products based on new learning, the German government, starting in 1887, built on the outskirts of Berlin the Imperial Institute of Physics and Technology (PTR), which at that time was the best equipped and most expensive laboratory in the world. The need to develop a better light bulb drove research on the radiation of heat and light from hot objects. New experimental results early in 1900 showed that the classical theory describing radiation was flawed (the first clue). Planck, who by this time was the senior physicist at Germany's (foremost) University of Berlin, decided to work on the problem.

Fig. 2.1. Max Planck. (Image from AIP Emilio Segre Visual Archives.)

Max Karl Ernst Ludwig Planck was born on April 23, 1858, and was brought up in a well-to-do and cultured family. His father, descended from clergymen, had become a professor of constitutional law in Kiel, which was then a part of Danish Holstein. Max was a good student, excelled as a pianist, and could have pursued a career in the arts, but he followed instead a curiosity for physics. Professors at the time of his schooling were state supported, but theoretical physics had not been established as a field of study. After doctoral studies with noted scientists at various universities and with no suitable position available, Max in 1880 began teaching as a privatdozent (a lecturer who was paid directly by his students and given a place to teach, but without a professorial position at the university). Eight years later, at the age of thirty, he was asked to succeed Gustav Kirchhoff in the now-established professorship for theoretical physics at the University of Berlin.

By late 1900 Planck had devised a mathematical expression that perfectly described the new experimental results on radiation, but it involved a wrenching rejection of classical ideas that he and everyone else held strongly. He was deeply troubled by the implications of what he had just found.

As background to understanding the strangeness of Planck's achievement and, as it turns out, much of quantum mechanics, I provide the following brief history and introduction to what was known at the time about light and heat radiation.

> Isaac Newton, in the mid sixteen hundreds, had theorized that light was composed of particles, which he called *corpuscles*. This view was overturned by a bold twenty-seven-year-old Thomas Young, who in 1801 dared to oppose the views of the great Newton. Young found that light of any particular color impinging on two slits in a barrier would undergo a process of diffraction and interference that occurs only with wavelike entities. I describe this process here by analogy to what happens in water.

Figure 2.2 illustrates the diffraction and interference of waves of water as they impinge on two openings in a wall with water on both sides. Note that waves do not really travel; the water molecules mainly move up and down: water rises to a peak at one position followed by water a little farther ahead rising to a peak moments later, much like the human wave produced by football fans in successive sections of a stadium when they raise then lower their arms. But unlike this human wave, which propagates around the stadium, the waves in our figure propagate from left to right.

Now suppose that the wave (of crest-to-crest wavelength w) impinges upon the wall as shown in (a). Part of each wave will pass through the two openings and emerge on the other side, diffracting (spreading outward) from each opening as shown in (b) at Time 2 and progressing as shown through Time 4. The circular set of diffracted waves emerging from one opening will interfere with the set emerging from the second opening. What results is an interference of crossing waves, illustrated by the crossing of the lines representing the crests of the waves at Time 3 and the interference pattern of wave heights shown in (c) at the position of the dashed line L at Time 4. Where crest crosses crest, the water rises to the combined height of both waves. Where trough crosses trough, the water sinks to the combined depth of both troughs. Where trough crosses crest, there tends to be a cancellation and the water does not rise or fall much, if at all.

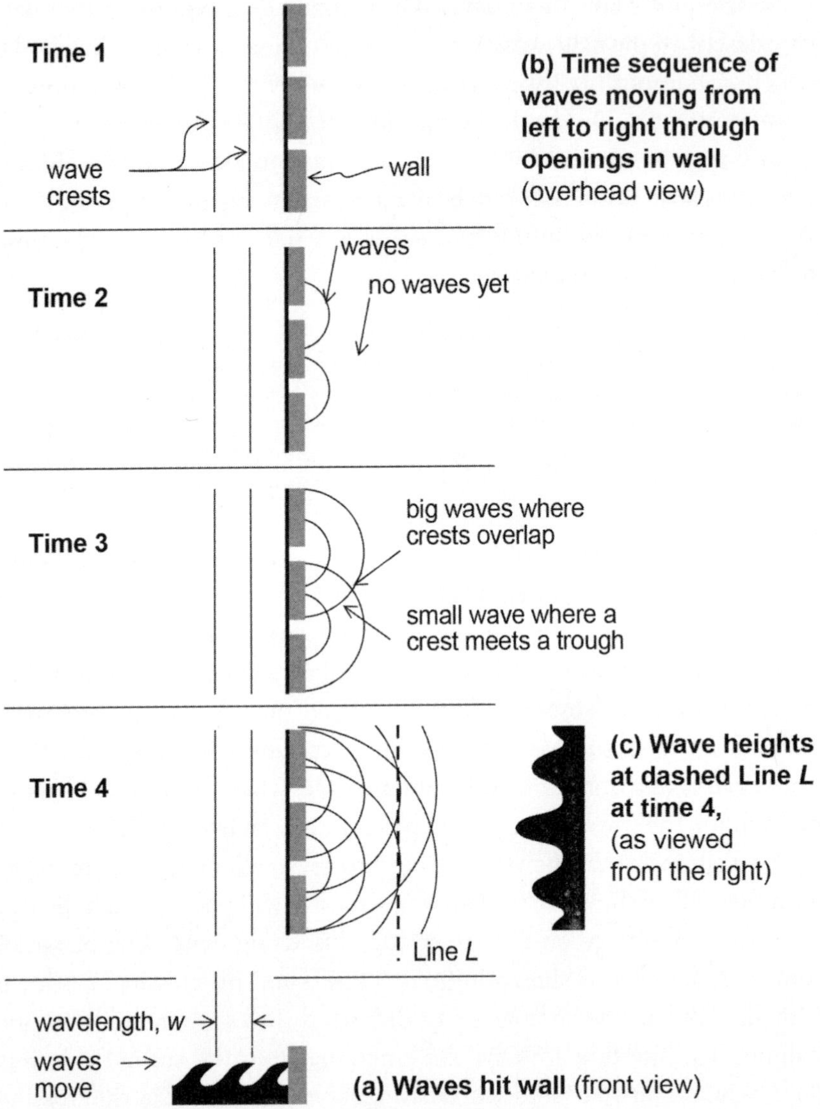

Fig. 2.2. Diffraction and interference of waves in water: (a) Waves of crest-to-crest wavelength *w* propagate from left to right, (b) diffract through two openings to interfere as described at Time 3, and (c) create the interference pattern of wave heights at line *L* as shown at the bottom right. (Modified, with permission, from Fig. 5.1 of *Big Bang, Black Holes, No Math*, by David Toback. Copyright © 2013 by David Toback.)

This was for water, but what about for light? The interference pattern shown at the right of this figure is similar to the pattern for light that was presented by Young to the Royal Society in London to show how light behaves. The argument was that no particle could pass through both slits and interfere with itself. And Young's experiments, measuring the intensity of light in the interference region, demonstrated just the sort of pattern that we see with the interference of waves in water.

It would seem that the conclusion had to be that light was composed of waves, not particles. But at the time, this conclusion was not accepted.

> According to Kumar, "When he first put forward the idea of interference and reported his early results in 1802, Young was viciously attacked in print for challenging Newton. He tried to defend himself by writing a pamphlet in which he let everyone know his feelings about Newton: 'But much as I venerate the name of Newton, I am not therefore to believe that he was infallible. I see, not with exultation, but with regret, that he was liable to err, and that his authority has, perhaps, sometimes even retarded the progress of science.'[1] Only a single copy was sold."[2]

By Planck's time, the situation had reversed, and it was hard to get anyone to believe that light was anything but a wave. All on the basis of sound theory and experiment. But, as you will soon see, new evidence to the contrary began to build.

In the next few paragraphs we examine how it is that waves are characterized. This may seem like detail, but it will be very useful later on.

Planck's analysis, one hundred years after Young's experiments, contradicted the firmly held classical view of light as a wave. To fit his theory to the measured radiation data, Planck required that light or heat radiation be released in discrete particlelike packets of energy, which he called *quanta*. And to make his analysis work, the energy of each quantum was defined as a constant (later called Planck's constant, represented by the symbol h) times the frequency of the light: in simple shorthand, $E_{\text{quantum}} = hf$. So, the lower the frequency (that is, the longer the wavelength of light, tending toward red in color and the infrared), the lower the energy of the associated quantum. And, conversely, the higher the frequency (that is, the shorter the wavelength of the light, tending toward the violet and ultraviolet), the higher the energy of each quantum. (As you will come to see, Planck's constant is as fundamental to physics as the number π is to geometry and mathematics.)

If you measure the distance crest-to-crest (the wavelength, *w*) and count how many wave crests pass by the wall in a given time, you have the wave frequency, *f*, in waves per second (cycles per second, where each cycle is the passage of a crest and a trough and then a crest again). Then you can calculate the speed of wave propagation, labeled here as *S*, by multiplying *w* times *f*. In scientific shorthand we write this as $S = wf$.

By 1900, light and heat radiation had come to be recognized as just two of many wavelength ranges of electromagnetic waves, the only difference between them being their wavelengths. Heat radiation is of longer wavelength than light, into the infrared, and therefore not seen. All electromagnetic waves propagate at "the speed of light," *c*, approximately 3×10^8 meters per second. So, for electromagnetic waves our formula is specifically $c = wf$. (Dividing both sides of this equation by either *w* or *f*, which still keeps it in balance, gives us $c/w = f$ and $c/f = w$, which are handy formulas for calculating either *f* or *w* if the other is known. If one knows *w*, divide it into *c* to get *w*. If one knows *f*, divide it into *c* to get *f*. Stated another way, if you specify either *w* or *f*, it's as if you have specified both properties. Sometimes it's more convenient to discuss an aspect of physics using *w*, at other times it's more convenient to use *f*.)

(A very good description of the classical view of electromagnetic waves is provided in Appendix A. If, for now, you'd just like to get a sense of the many types of waves and rays that compose the electromagnetic spectrum, you might only glance at Figure A.1(c) and then view the spectrum of wavelengths described at the left in Figure A.2.)

Nothing in Planck's classical understanding of physics would allow light and heat to be radiated in discrete chunks of energy. Light had been universally thought of as being wavelike, with any amount or intensity possible. That is, light in Planck's earlier experience could be made dimmer and dimmer without limit. At any frequency it would be available in any amount of energy, with no minimum. Now, to fit the radiation data, he had postulated a light quantum. How would this (particlelike?) indivisible quantum diffract through two slits and interfere with itself, as Young had demonstrated?! It seemed impossible! And Planck indeed believed that it *was* impossible. (He was not about to return to Newton's ideas of corpuscles, which had been shown to be wrong!)

Nevertheless, Planck presented his results, passing off the apparent contradiction between the quantum and classical ideas as something to be resolved later on. For more than ten years afterward, Planck labored to achieve that resolution but couldn't do it. He gave his quanta physical basis, imagining that the light was emitted by a collection of microscopic electric oscillators operating at different frequencies and housed on the surface of the emitting hot materials. But, to fit experimental results, he was always stuck with having to make the assumption of discrete chunks of energy. He wouldn't go so far as to say that light itself was quantized, only that the transfer of energy between light and substances occurred in quantum units. And he resisted for the rest of his life the revolution in physics that his idea of the quantum produced. But he was right in what he had presented in 1900, and in 1918 he was awarded the Nobel Prize in Physics[3] *"in recognition of the services he rendered in the advancement of Physics by his discovery of energy quanta."*

THE PHOTOELECTRIC EFFECT (THE SECOND CLUE)

The beginnings of a radical break away from classical physics came in 1905 with Einstein's analysis of the photoelectric effect. With it he began to transform the quantum from Planck's item of mathematical convenience to the basis of an entirely new concept of physics that would eventually (but not soon) gain acceptance.

In the photoelectric effect, light shining on a metal, under the right conditions, causes electrons to pop out of the metal's surface, as sketched in Figure 2.3. No matter how intensely the light might shine, unless the wavelength of the light was shorter than some particular value characteristic of each metal (so that the energy of the light quantum was greater than a corresponding particular value), no electrons would be knocked free from the metal. Yet some electrons would pop free even with very little light radiation, as long as the wavelength of the light was short enough (energy of the quantum high enough). There was no classical explanation for these results!

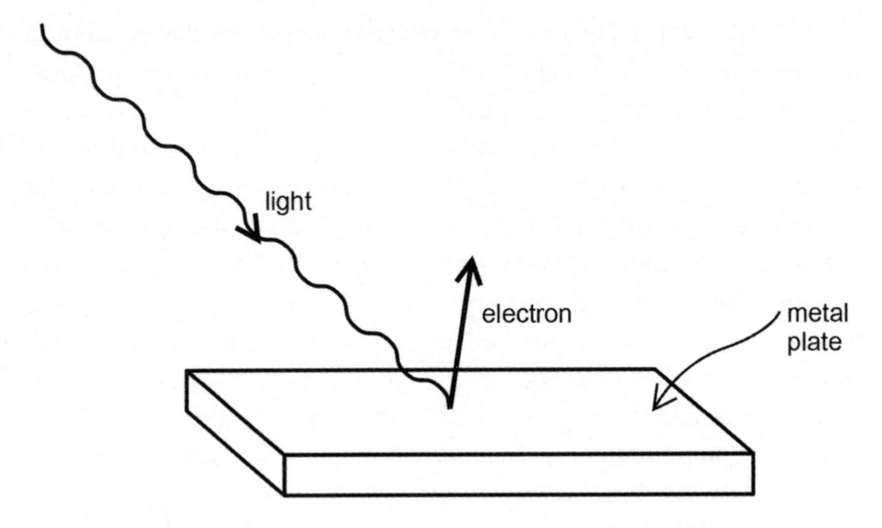

Fig. 2.3. The photoelectric effect: Light of sufficiently short wavelength (sufficiently high photon energy) knocks an electron free from a metal plate.

We'll describe Einstein's analysis of this effect shortly. But first, realize that Einstein was just twenty-six years old at the time of this analysis, and he was supporting his wife and young son by working as a patent examiner for the Swiss government. We take a moment now to relate the twists and turns in Einstein's early life that may have prepared him to tackle this analysis and prepare him to create some of the most profound and important ideas and explanations of modern physics.

Albert Einstein was born on March 4, 1879, in Ulm, Germany, into a secular Jewish family. His sister, Maja, came along two years later. It took him a long time to learn to speak, and when he did so it was after first constructing and examining sentences in his head. It was not until the age of seven that he began to speak normally.

Albert's interest in the workings of the physical world is said to have begun at the age of five when he became intrigued by the unseen forces that moved the needle of a compass. By the time he was six, his family had moved to Munich, where his father and uncle opened an electrical business. And so he was exposed to electrical machines and concepts of electricity and magnetism. As a boy he developed a preference for doing things by himself, demonstrating much patience and tenacity in solitary pursuits; for example, at the age of ten, he built card houses to as high as fourteen stories.

As a preteen, Einstein had become enamored of religious studies. But then he totally rebelled at what he considered deception after being exposed to the ideas of Euclid, Kant, Spinoza, and the *Popular Books on Natural Science* by Aaron Bern-stein. This through the visits of Max Talmud, a penniless twenty-one-year-old Polish medical student, who every Thursday was invited to join the Einstein family for dinner.

The family business manufacturing direct-current dynamos and meters initially prospered, to the point that the brothers were installing power and lighting networks, including the lighting for the 1885 Munich Oktoberfest. But they eventually lost out to larger companies, particularly to Siemans's alternating current systems, and in 1894 the brothers relocated the business to Milan. Albert, now fifteen, was left behind with distant relatives to complete the remaining three years of secondary school. But he was worried about compulsory military service in militaristic Germany if he became seventeen as a German citizen. (Though he would later become a pacifist, it was not service, but the militarism of Germany that he hated.)

Albert found an excuse to leave school, joined his family in Milan, and renounced his German citizenship. On finally[4] passing entrance exams to the Swiss Polytechnic School (ETH) in Zurich, he elected a course that would qualify him to be a schoolteacher of math and physics. He was the youngest of six such students entering in October 1896. Eventually he fell in love with one of them, a Hungarian Serb of good family, Mileva Marić, who in January 1903 would become his wife.

Things were at first good at school, but then came some very difficult times.

Einstein enthusiastically explored interesting questions in physics with Mileva and friends, but he managed to antagonize his professors by doing things his own way. He skipped classes, borrowed notes to study, and barely graduated. Mileva failed final exams and returned home. Einstein, without good recommendations from his professors, could not get a job. Unlike many of his classmates, he was not offered an assistantship at the ETH. And both of their families were opposed to the couple getting married, especially Einstein's mother.

Albert eventually managed to get into a doctoral program at the prestigious University of Zurich, but by then his father's business was failing. Albert earned some money tutoring. Mileva had become pregnant, and she failed her exams, again. In a difficult delivery at home, she gave birth to a baby girl whom they named Lieserl. Albert wrote loving and supportive letters, but he had no money to travel. He would never see his daughter, who is presumed to have been given up for adoption. He hid in pursuit of his physics.

Finally, in June 1902, with the help of friends, Albert found a job, starting work at the Swiss patent office as technical expert third class, but with a pay twice what he would have made as an assistant at the ETH. Now, with a secure income, he obtained his father's (deathbed) permission and married Mileva at the Bern City Hall in the presence of two of friends. He and Mileva are shown some years later in Figure 2.4. In May 1904, they began a family anew, when their first son, Hans Albert, was born.

Fig. 2.4. Albert Einstein and his wife, Mileva Marić, in 1912. (Image from ETH-Bibliothek Zurich, Bildarchiv.)

Einstein's work at the patent office required that he be skeptical of all that he read. This perhaps gave him an uncommon grounding for seeking out flaws in classical physics. And the atmosphere of the patent office served well for critical thinking. Though he worked a six-day, forty-eight-hour week, he was able to use the time at his desk to both do well at his patent work and, without interference, explore the fundamental laws of physics. But it wasn't just the quiet of the patent office that allowed him to function. Einstein showed tremendous powers of concentration, and he was able to work amidst the chaos of a busy home or withdraw into his thoughts in the middle of a social gathering.

(This brief biography has necessarily been limited. To get a better feel for Einstein, his struggles, triumphs, virtues, and failings, I highly recommend that you read from page 102 of *The Human Side of Science*, by Arthur Wiggins and Charles Wynn [Reference KK]. Complementary observations are offered in *Quantum*, by Manjit Kumar [Reference K].)

Now, as further background to Einstein's analysis of the photoelectric effect, I describe what in 1905 was understood of the electron:

The discovery of the electron in 1897 started with the observation of what had been called "cathode rays." If a large voltage was created between two wires that fed into a partially evacuated glass tube, electricity would flow through the gap between the wires. This would cause the small amount of air still trapped in the tube to glow, marking the path of what were called "rays." By studying these rays, J. J. Thomson in 1897 determined that they were actually streams of small particles of negative charge, each less than one-thousandth the mass of the hydrogen atom. Applying the label that Newton had used much earlier for particles of light, he called the particles *corpuscles*, but they later came to be called *electrons*.

Thomson's illustration of the glass-enclosed apparatus that he used for his experiments is shown in Figure 2.5(a).

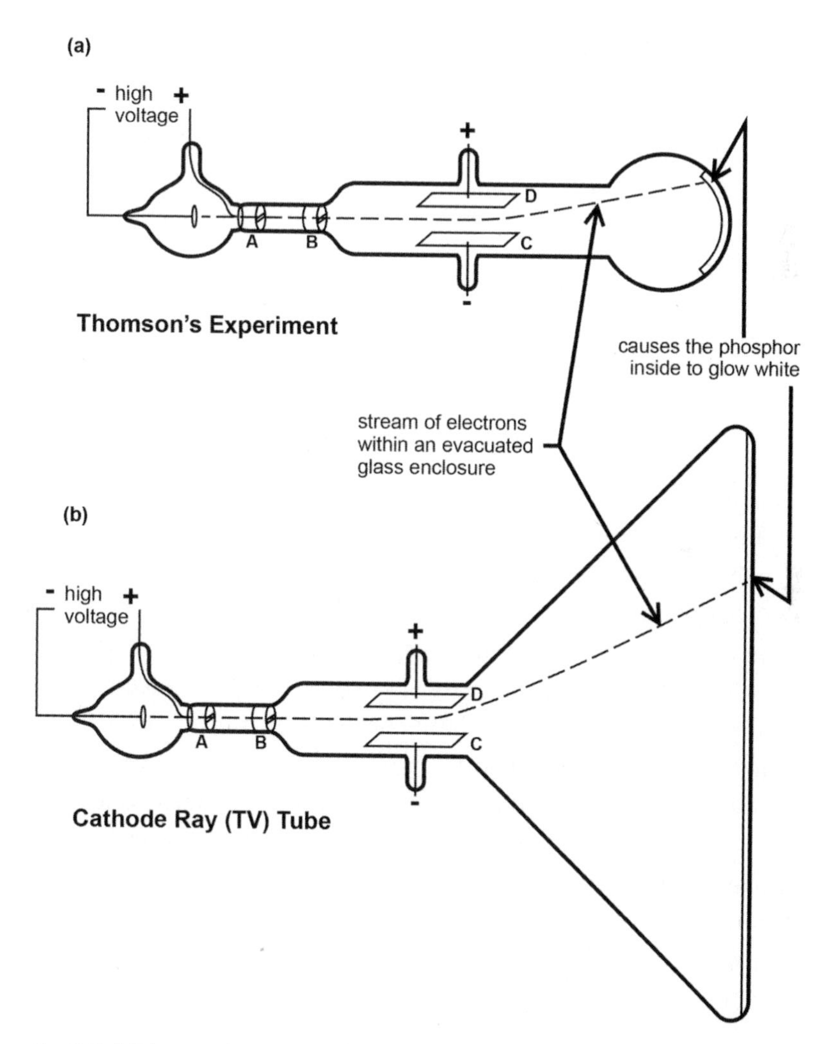

Fig. 2.5. (a)Thomson's apparatus for discovery of the electron. An evacuated glass enclosure houses (at left) a hot cathode at negative voltage from which is drawn a stream of electrons (dashed line) to and through the + voltage ring and deflected by + and – voltage plates at D and C to hit and glow a phosphor screen at the right. (b) Concept of an early type of cathode ray TV tube.

In his experiment, electrons (which always have a unit negative charge) are attracted and accelerated in vacuum within a glass enclosure from the metal cathode (C) toward the highly positive voltage brought by a wire to the region of the slits (A) and (B) that would allow only a ribbon-shaped beam of particles to pass. Those electrons passing through the slits are carried by their momentum to impact the phosphor on the inside of the front of the tube at the far right, causing it to glow. By applying smaller + or – voltages across the plates (D) and (E) the electrons (each traveling a path like that shown by the dashed line) would be deflected upward or downward. Because the charge on the electron is negative, it would be attracted to the positive plate and repelled from the negative one (which is what happened).

Because electromagnetic waves are not expected to be deflected by electric fields at all, Thomson was able to conclude that the rays observed were caused by a charged particle. He correctly assumed that this was the smallest charge to be found in nature, and it would be assumed to be nature's basic unit of charge. From the deflection and the voltage causing it, he was able to calculate the ratio of the electron charge to its mass. Comparing this ratio to the ratio of charge to mass for the hydrogen nucleus (later found to consist of one proton of positive unit charge), Thomson was able to roughly estimate the electron's mass (actually about 1/1,800 the mass of the proton).

Einstein followed after Planck in examining theoretically the radiation of heat and light from hot bodies. But he used a statistical approach and a different model. He was able to derive Planck's formula for the energy of the quantum. And he concluded that it was the light itself that was quantized, not the hypothetical oscillators in Planck's model. He applied this concept of light quanta to explain the photoelectric effect.

Einstein reasoned that the light consisted of many quanta of energy, and that even one single quantum of light, if it had a short-enough wavelength (that is, if it had sufficient energy), would pop loose one electron as shown in Figure 2.3. In that sense, the quantum might be viewed as a "particle" of light. One light particle of sufficiently high energy, that is, of sufficiently short wavelength, pops loose just one electron.

Now I digress briefly to provide a glimpse of a couple of the products that resulted from Thomson's science.

It was eventually found—in more completely evacuated tubes of other geometries—that electrically heating the cathode to the point where it would glow red-hot would reduce the voltage needed to induce electron flow. And, without the deflecting plates, slits, and phosphor, it was found that a very small voltage across an additional wire between the cathode (C) and anode (A) could cause large changes in the flow. These "tubes" would allow the amplification of small electrical signals and the invention of a host of devices that we now label as "electronics." Much later still, in the 1950s, it was found that amplification could be obtained in tiny composites of solid materials. This introduced "solid-state electronics," quantum devices eliminating most* of the need for the cumbersome tubes and eventually leading to the production of integrated circuits and devices within single "chips," as described briefly in Chapters 17 and 23.

(*Among the holdouts were the large, heavy "cathode-ray tubes" that until recently preceded modern flat-screen TVs. One such tube is shown schematically in Figure 2.5(b) as a much larger phosphor-screen version of Thomson's apparatus. A spaghetti-thin stream of electrons in this tube is directed by sideways or vertical plates and voltages to impact the front of the tube at various locations, causing a glow of the phosphor that is seen through the glass from the outside of the tube. By changing plate voltages to cause the stream to scan rapidly in lines across the front of the tube, while at the same time turning the electron flow on and off, light and dark regions are produced to make a picture that persists momentarily and is then quickly changed by the immediately following scan. In this way, small changes in the picture are made to occur rapidly and seamlessly to produce what we view on TV as motion. This is basically what occurs in older black-and-white TVs. Further innovations produced color.)

The quantum now explained two phenomena in defiance of the laws of classical physics. And Einstein, unlike Planck, would champion the quantum view. But there were still many questions that needed answering.

Because of diffraction and interference, light was considered to be wavelike. But now, with the photoelectric effect, light was also shown to behave like a bunch of quantum particles. (Back to "corpuscles"?) How does one get interference from a particlelike light quantum? How does a light quantum, an indivisible chunk of energy, go through one slit (with reference to the water analogy of Figure 2.2), and then interfere with itself once it has passed through the slit, to form the interference pattern that Young had observed?!

"IF THIS IS CORRECT, THEN EVERYTHING WE THOUGHT WAS A WAVE IS REALLY A PARTICLE, AND EVERYTHING WE THOUGHT WAS A PARTICLE IS REALLY A WAVE."

In a lecture in 1909 before an audience of the leading German physicists of that time, Einstein presented a mathematical model that suggested that light would have the properties of *both a particle and a wave*, what would later come to be called *wave-particle duality*. Planck, who chaired the session, politely thanked him but disagreed, stating what most physicists then still believed: that quanta were only necessary in considering the exchange of energy between matter and radiation, and that it wasn't necessary to treat light itself as a particle or actually made up of quanta.

Einstein's paper on the photoelectric effect, his paper on special relativity, and a third paper on Brownian motion (as evidence for the existence of atoms and molecules) were among five papers that he published in his 1905 "miracle year." Based on this and much subsequent work, Einstein would in 1921 be awarded the Nobel Prize in Physics *"for his services to Theoretical Physics, and especially for his discovery of the law of the photoelectric effect."*

A QUANTUM THEORY OF SOLIDS (MORE EVIDENCE)

In 1909, after a prerequisite year working as a privatdozent at the University of Bern, seven years after he'd joined the patent office, Einstein finally acquired a position equivalent to assistant professor of theoretical physics at the University of Zurich. There he showed a relaxed teaching style, encouraged questions, and soon gained the respect of his students. And he turned his attention to a completely new area of physics.

Many solid substances consist of crystals of atoms stacked in three-dimensional arrays. What we sense as temperature is the heat-induced jostling of atoms about their crystalline lattice positions. The more extreme the jostling, the higher the temperature. (When we touch a hot object, that jostling is transferred to the atoms in our fingers. If the jostling is extreme, the cells in our fingers are damaged and burns result.) If a solid like a metal gets hot enough, the jostling becomes so extreme that the atoms lose track of their lattice positions and wander around, causing the solid to melt.

Einstein assumed that the jostling could be modeled as a superposition of mechanical oscillators of different frequencies, analogous to weights on springs. As Planck had done for light radiation, Einstein also assumed that only certain quantized frequencies would be available for the oscillations. He derived a formula for the manner in which temperature would rise as heat was added to the solid. That is, he calculated what is called the solid's *heat capacity*.

For two years, Einstein's theory received little attention. But then Walter Nernst, an eminent physicist at the University of Berlin, succeeded in accurately measuring the heat capacities of solids at low temperatures. The results exactly matched the predictions of Einstein's calculations!

Einstein was now beginning to be noticed. He was offered, and he accepted, a professorship at the German University in Prague; and he moved there in 1911 with Mileva and his sons Hans Albert, now six, and Eduard, almost one. He was also invited to the First Solvay Conference, held in Brussels, to be the prestigious last speaker of eight who would make presentations before twenty-two of the leading physicists across Europe. The conference addressed questions regarding molecular and kinetic theories (i.e., theories about the motion of particles). He would speak on the heat capacity of solids. It was the first such conference to involve a quantum concept (but light quanta were not on the agenda).

Einstein had succeeded admirably, but the years from 1908 through 1911 were especially hard on Mileva. While privatdozent (unsalaried lecturer) at the University of Zurich, Albert still worked his full-time job at the patent office. When a year later he acquired his assistant professorship, he had a heavy teaching load, and he became popular with the students who would surround and follow him to the cafés in Zurich to discuss physics. In July 1910, Mileva gave birth to their second child, Eduard, who unlike Hans Albert was very fussy. Mileva looked after two children, and she felt neglected by a very busy husband. There were also misunderstandings. And later she was very unhappy with living in Prague. Albert also longed to return to Zurich. With the aid of Marcel Grossmann (fellow student and friend, now dean of mathematics and physics), he obtained an appointment as a master physicist to the renamed Swiss Federal Technical University (ETH), where he had earlier been unable to find a job even as an assistant.

THE QUASI-CLASSICAL QUANTIZED BOHR MODEL OF THE ATOM (A SOLUTION OF SORTS)

A major breakthrough came in 1911 from the Danish physicist Niels Bohr, seen to the far right of the middle row in Figure 1.1. Bohr would become the driving force in integrating the various experimental and theoretical contributions of the early quantum theory to provide a description of the atom and the elements. Later, through his Institute for Theoretical Physics and his insights and communications with other leaders in the quantum field, he would become the foremost proponent and spokesman for the Copenhagen interpretation of the developing quantum mechanics.

To understand the magnitude of Bohr's accomplishment, we first provide a sketch of what was known of the atom at the time.

In 1903, just six years after his discovery of the electron, J. J. Thomson proposed what was called the "plum pudding" model of the atom. His model assumed that the atom of any element would be composed of thousands of (negatively charged) electrons embedded in a ball of massless, spread-out, comparable positive charge. Later he would revise his model to include fewer electrons and assume that most of the mass was in the region of positive charge. At the time, many eminent physicists and chemists still believed that there were no such things as "atoms."

Ernest Rutherford, soon to become Bohr's mentor, would develop a different model for the atom.

Rutherford was one of twelve children raised in a working-class family in New Zealand. Through a series of scholarships Rutherford had come to Cambridge in 1895 to study under Thomson, initially continuing work that he'd begun earlier to devise ways of detecting radio waves, and then examining the radiation emanating from radioactive uranium. With Thomson's high recommendation, he was appointed in 1898 to professor at McGill University in Montreal. There he worked with radioactive elements. In 1901 with fellow professor Frederick Soddy he discovered that one radioactive element could transform into another through the radiation of alpha particles, later recognized as helium nuclei. (It was Rutherford who coined the term *half-life* to describe the time over which an element would lose half of its radioactivity.) For this work he would in 1908 be recognized with a Nobel Prize in Chemistry and the offer of a promotion to professorship at the University of Manchester. Soddy would get the prize two years later.

At Manchester, Rutherford assigned his assistant Hans Geiger (yes, of the Geiger counter) and an undergraduate, Ernest Marsden, to examine the constituency of the atom by bombarding metallic foils with helium nuclei. When these heavy particles sometimes bounced back rather than penetrating through, Rutherford concluded that the atom had a yet heavier nucleus, which his calculations showed to be 100,000 times smaller than the atom. He devised a "planetary" model of the atom, of tiny electrons in large orbits circling this tiny but heavy nucleus, to create an atom of the overall size inferred from experimental measurements. But there was one major, major flaw in his model (to be described shortly). He decided to ignore it, and at a meeting in Manchester in March 1911 he put forth his ideas. Because of this flaw, the reception was cool, and later that year, at the First Solvay Conference, his planetary model wasn't even discussed.

The problem was that the planetary model ran counter to a demonstrated and well-known characteristic of charged particles: if they are accelerated, they radiate energy. Like the planets, the electron in its orbit undergoes acceleration, but because the electron has net electrical charge, Rutherford's electron in orbit would be expected to radiate energy, slow down, and in some manner collapse. The atom would not be stable. And this, of course, was not (and is not) observed.

The Acceleration of an Object in Orbit

By looking sideways at planets or electrons in hypothetical circular orbits, as shown in Figure 2.6(b), one can see that the electron accelerates. Seen this way, the electron (or planet) would appear to be oscillating back and forth, from left to right and back again, accelerating first in one direction and then in the other. Even viewed from the top, as shown in Figure 2.6(a), the electron (or planet) is seen always to be accelerating toward the nucleus (or the sun) in its orbit; otherwise, its momentum would just carry it off to infinity in a straight line.

According to everything that we know (and that was known at the time) accelerating electrons, or any accelerating charged objects, will continuously broadcast out electromagnetic energy. (That is how radio waves are produced: by accelerating billions of electrons back and forth or up and down at the frequency at which transmission is desired. But, unlike the electron in an atomic orbit, electrons in a broadcast antenna are continuously resupplied with energy at a rate of so many kilowatts of broadcast power, as is often advertised.) So, according to what was known in the physics community, the electron in Rutherford's orbit would be broadcasting out energy. This would cause the electron to spiral down and collapse the atom. His model would not be viewed as valid.

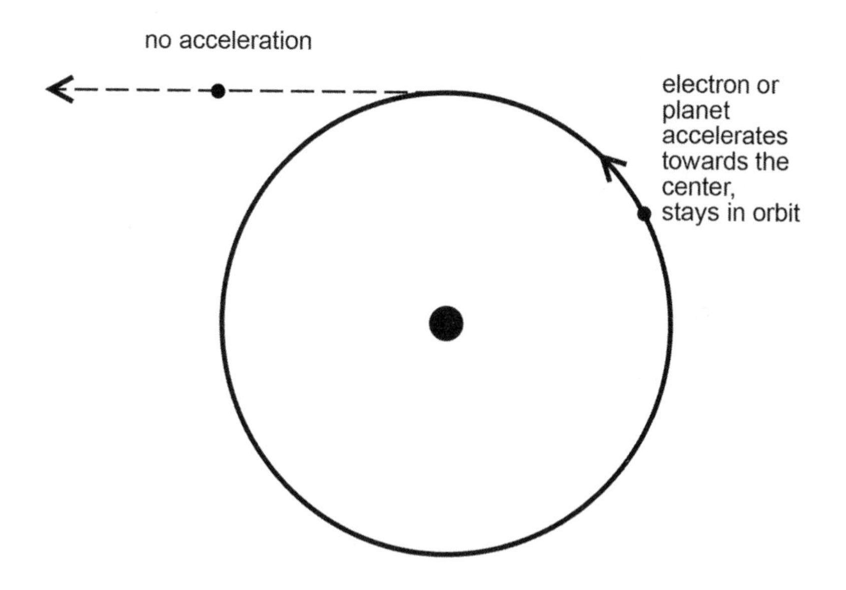

(b) Side View

Fig. 2.6. (a) A planet in orbit around the sun follows a circular or elliptical (not shown) path because the gravitation of the sun's mass continually attracts it (accelerates it) toward the sun. Otherwise, the planet would travel in a straight line (shown dashed). An electron, attracted to the positive charge of a proton, will behave in a similar manner. (b) In this side view, the planet or electron would appear to be accelerating back and forth (right and left).

Bohr, like Einstein, had a nose for the importance of areas of physics showing problems and contradiction. He had become convinced that Rutherford's model of the atom was basically right, and he decided that the answer to the problem of radiation must somehow be connected to the (still mainly disbelieved) ideas of Planck and Einstein on the quantum.

Niels Henrik David Bohr was born on October 7, 1885. He enjoyed a privileged childhood in Copenhagen: his mother's family was wealthy in banking and influential in politics; and his father was the distinguished professor of physiology at Copenhagen University. Niels and his older sister and younger brother were exposed to intellectual discussions during regular visits of writers, artists, and scholars of all sorts to the Bohr home. Niels was athletic and excelled in mathematics and science. In 1903 he enrolled to study physics at Copenhagen University, receiving a master's degree in 1909 and a doctorate in 1911, with a dissertation on the theory of metals.

In September Bohr left on a one-year postdoctoral scholarship to work with the (by then) Nobel laureate J. J. Thomson at Cambridge. But there he found it difficult to get attention and communication from Thomson, and after eight months he arranged instead to join Rutherford in Manchester for the rest of his scholarship year, despite the fact that Rutherford was doubtful of theorists. (That Bohr played soccer apparently helped him to fit in once he got there.)

Following his instincts, Bohr postulated that the electron in an atom could only be found in certain discrete circular "stationary" orbits, and that the electrons would be stable in these orbits and not radiate energy as classical physics required. He had no proof, no justification, except, eventually, that it would work and that he could use his model to quantitatively explain much as-yet-unexplained phenomena.

In support of his postulate, Bohr fastened on an idea published earlier that year by J. W. Nicholson, a former colleague at Cambridge: that the momentum (mass times velocity, $M \times v$) of a ring of electrons, times the radius of the ring, would be quantized in integral units (labeled n) times Planck's constant divided by 2π. In physics shorthand, we write this as $Mvr = (nh)/(2\pi)$. He showed that if this formula were to hold true and only one electron per orbit was considered, the mathematics of classical Newtonian physics would allow orbits of only certain sizes proportional to n^2. The first five of these possible orbits are shown for the one-electron hydrogen atom in Figure 2.7(a). The same mathematics showed that the electron, if it were to be found in one of those orbits, would have a corresponding distinct, precise energy, as shown for some of the orbits in Figure 2.7(b). (This is quite unlike the planets, each of which would seem to take on any size of orbit and any of an associated continuum of energies. Each planet just happens to be in one particular orbit. With just a little more or less energy, each could easily be in another nearby orbit.)

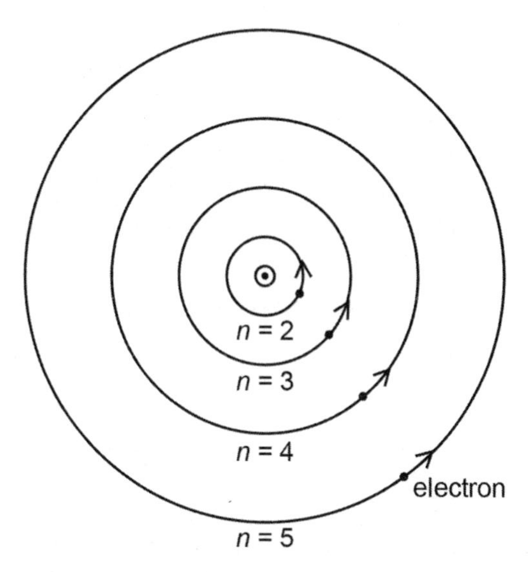

(a) Bohr orbits for the electron in the hydrogen atom

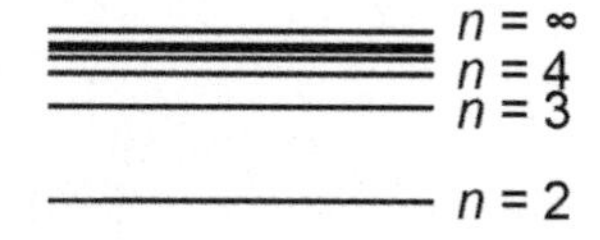

(b) Corresponding Bohr orbit energy levels

Fig. 2.7. (a) Relative sizes of the five lowest-energy Bohr orbits for the electron around the proton (center dot) in the Bohr model of the hydrogen atom. (The electron and proton would be too small to be seen, even at this approximately 100-million-times magnification.) (b) Note that the $n = 1$ is the lowest energy level and corresponds with the smallest orbit [label not provided in (a)]. (Energy levels will be defined in Chapter 11.)

Bohr's $n = 1$ state, with the smallest orbit and lowest energy, would be called the *ground state*. Since no smaller stable state existed, and (by postulate) the electron could only be in these stationary stable states, by postulate the atom could not collapse as classical theory would expect. In this way, Bohr would explain the observed sizes of atoms.

The Bohr model gained credibility when it exactly predicted the results of experimental data that had defied explanation for nearly sixty years. To understand the significance of this, we need to learn a little bit about spectra.

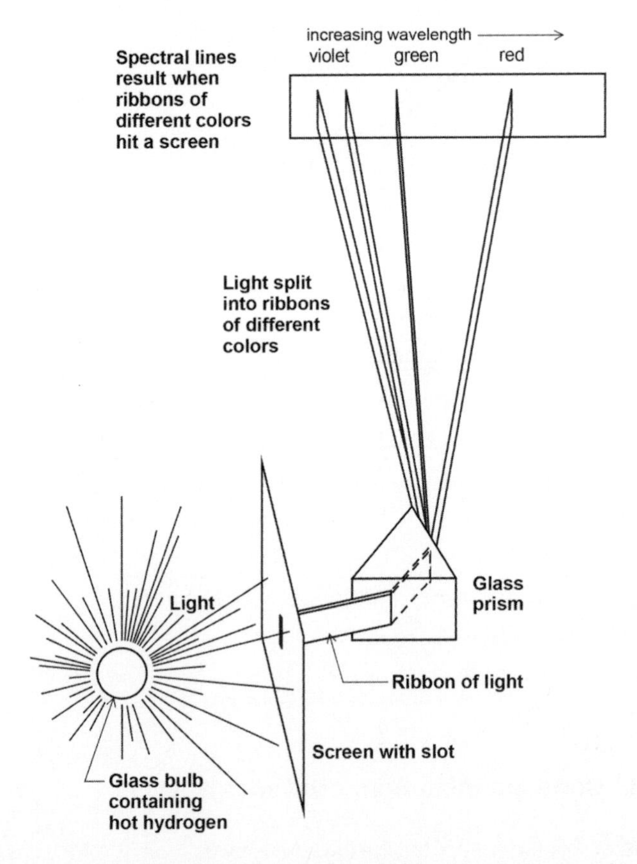

Fig. 2.8. How a spectrograph works. Light emitted from hot hydrogen is bent at different angles through a glass prism, depending upon the light's wavelength. A ribbon of light passing through a slot in a screen is shown split into four ribbons, each with a color (wavelength) characteristic of a light quantum (photon) of that wavelength emitted during an electron's change of orbit (energy).

Isaac Newton, experimenting as early as 1666, caused white light to split into a rainbow of colors (which he called a *spectrum*) as he passed the light through a prism of glass (glass shaped to a triangular cross section). In the early 1800s it was learned that various substances, either sparked or exposed to flames, would emit light in a mixture of different colors, and that the individual colors from these substances could be separated using prisms in spectrometers. In the spectrograph sketched in Figure 2.8, separated ribbons of light emitted from hot hydrogen show as bright-colored lines on an otherwise-dark background. Each element or compound displays its own characteristic set of colored lines.

Systematic study of spectra to identify elements and compounds began in the 1860s with the collaboration of the chemist Robert Bunsen at Heidelberg and physicist Gustav Kirchhoff (remember—Planck's predecessor at the University of Berlin). To provide a clean flame hot enough to excite the spectral emissions, Bunsen and university mechanic Peter Desaga developed what came to be known as the Bunsen burner (which, as Sam Kean would phrase it, "made him a hero to everyone who ever melted a ruler or started a pencil on fire" in chemistry lab[5]). Studying under Bunsen as a graduate student at about that time was Dmitri Mendeleev, a young Russian whose subsequent work would figure prominently in the characterization of the elements and the development of the periodic table of the elements. (I'll say more about Mendeleev and the development of the periodic table in Appendix B.)

HYDROGEN LINES (SOME LONGSTANDING CLUES)

It was the lines of the emission spectrum of hydrogen shown in Figure 2.8 that clinched Bohr's model. The hydrogen lines had been measured as early as the mid-1850s by the Swedish physicist Anders Ångström. At first glance, the hydrogen spectrum (turned vertical) might seem to resemble the lines showing the energy levels of the Bohr atom in Figure 2.7(b), but there is much more to it than that.

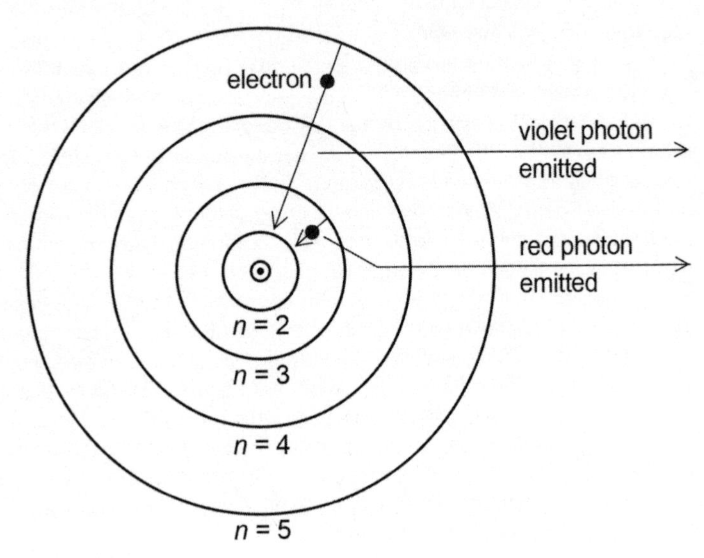

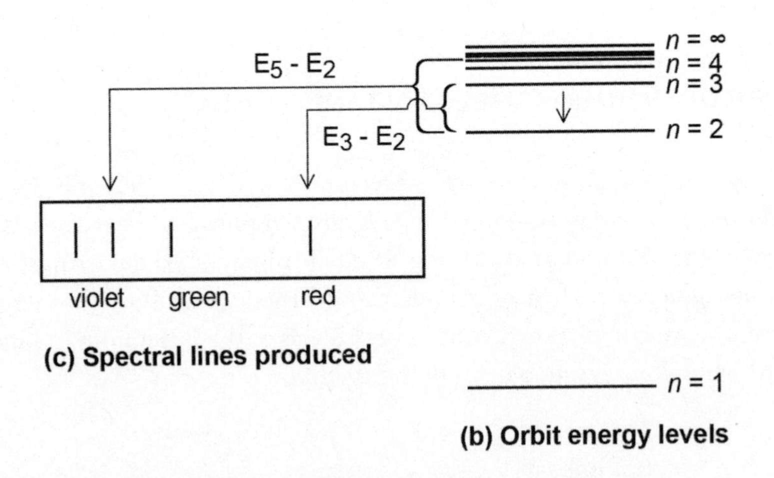

Fig. 2.9. (a) Transition of an electron from an $n = 5$ or $n = 3$ Bohr orbit to an $n = 2$ Bohr orbit, with the release of a violet or red wavelength photon. (b) Brackets showing the energy-level changes, that is, the energies released as photons during these transitions. (c) Violet and red spectral lines associated with these transitions, their photons, and the photons emitted.

An empirical formula describing the spectral lines in terms of integers had been devised in 1884 by a Swiss mathematics teacher, Jacob Balmer. Nothing in classical physics would explain the spectral lines or Balmer's formula. But when Bohr examined the formula, he immediately saw that his model would predict the formula and all of the observed spectral lines. Each line would be produced by the release of a light quantum in the transition of an electron from one of the higher of Bohr's energy states into a state of lower energy. The energy lost by the electron in each transition would be carried away as a single light quantum, as shown by arrows for the $n = 5$ to $n = 2$ and $n = 3$ to $n = 2$ transitions in Figure 2.9(a), with the energy lost in each case shown by the span of the brackets between energy states in (b). The spectral lines produced are shown in (c).

Bohr, like most scientists at the time, did not believe in the Planck-Einstein quantization of light, but he used Planck's formula $E_{quantum} = hf$ to calculate the frequency of the light wave that would be emitted for each transition. He would then be able to calculate the corresponding wavelength using the formula for electromagnetic waves that we derived earlier: $c = wf$ (which dividing by f would give $c/f = w$). He found an exact match between his calculated wavelengths and the observed wavelengths for all of the lines that were seen.

Bohr had explained the line spectrum of hydrogen as light carrying away energy to allow the transition of an electron from a state of higher energy to a state of lower energy. This could only result if he had a valid model for the atom and the energy levels of the states! (That he predicted some transitions that weren't seen would be a detail to resolve later.)

On March 6, 1913, Bohr sent Rutherford the first of three papers "On the Constitution of Atoms and Molecules" (as was the custom for review and forwarding by a more-senior scientist to get more-rapid publication). It was published in April. The second and third papers on the possible arrangements of electrons in the atoms of various elements were published in September and November.

Bohr's quasi-quantum, quasi-classical atom got a mixed reception when it was discussed at a meeting in England in September. On the Continent it was met with disbelief. But Bohr extended the theory to analyze spectral lines in light from the sun that didn't fit the hydrogen pattern, showing that these lines would roughly fit a Bohr model for helium (and this convinced the [by then] much-respected Einstein that Bohr's model had merit).

THE GREAT WAR AND ITS AFTERMATH—LIGHT QUANTA AND GENERAL RELATIVITY

World War I, which began on August 14, 1914, drained the laboratories of Europe in unreasonable nationalism that found friends and scientists supporting and fighting on opposite sides.

> Planck, now rector of the University of Berlin, sent his students off in support of "a just war," and signed a manifesto with ninety-three other luminaries asserting in defiance of the facts that Germany hadn't violated Belgian neutrality, had been forced into war, and had committed no atrocities. (He quickly regretted it and began apologizing to friends in other countries.) As a Swiss citizen, Einstein was not asked to sign the manifesto, but he was deeply concerned about it, and he helped to produce a counter manifesto against war and castigating German intellectuals for their blind nationalism. He was one of only four signers.
>
> Einstein had returned to Germany earlier that year, despite his aversions and in response to a most appealing triple prestigious offer delivered by Planck and Nernst: he would be appointed to one of the two salaried positions in the Prussian Academy of Sciences, to a unique professorship at the University of Berlin without any teaching duties, and to directorship of the soon-to-be-established Institute of Theoretical Physics. However, Mileva disliked the prospect of return to Germany. She had suspicions related to Alberts's increasing friendship with a divorced cousin Elsa in Berlin. And his mother was there. Mileva and Albert argued. Eventually they agreed on a separation, and Mileva returned with the two boys to Zurich.

Motivated by the demonstration in April 1914 (by James Franck and Gustav Hertz) of Bohr-like atomic transitions in mercury vapor under the bombardment of electrons, Einstein set about describing theoretically the mechanisms by which these transitions might take place. He found once again that the energy of light itself should indeed be quantized, and further that the quanta would have momentum and an associated direction of travel. (No mass, but momentum? Interesting!) All of this would seem to confirm light as a quantized particle. But acceptance of this idea would still not come until nine years later, when another experiment brought results that could only be explained if quanta were particles with momentum. (I'll describe that experiment early in Chapter 3.)

It was also during these years of World War I that Einstein completed his much-celebrated masterpiece, the theory of general relativity. The theory required the warping of space and time under the gravitational influence of massive bodies. It explained a precession of the planet Mercury's elliptical orbit around the sun and predicted a bending of starlight as it passed near the sun, which was observed during the solar eclipse of May 29, 1919.

With his theory of general relativity confirmed, Einstein's picture was on the front pages of newspapers, relativity was discussed in the streets, and Germany heralded a triumph of German science. He had in February finally been granted a divorce from Mileva, after promising her increased payments, his widow's pension, and the money he would receive from an impending Nobel Prize. But Germany was broken by war, with its people already hungry and soon to be laden with unreasonable financial burdens in reparations. By 1923, inflation would reduce the value of the German currency to an unbelievable exchange rate of four trillion marks to the dollar. Eighty billion marks would buy a loaf of bread. Even by 1919, in the postwar environment, anti-Semitism had begun to rise to the surface. Einstein, who had become a public and antiwar figure, was threatened even during his lectures, and, despite assurances from the government, feared for his safety. He had begun to shun public exposure. Two German Nobel laureate physicists (Philipp Lenard, who had measured the photoelectric effect, and Johannes Stark, who found that an electric field would split the spectral lines of hydrogen into closely spaced lines) had by now become rabid anti-Semites and promoted a group of scientists that in 1920 specifically and publicly denounced Einstein and attacked relativity as "Jewish physics." Nernst and others wrote articles to the newspapers in Einstein's defense.

EXTENSION OF THE BOHR MODEL (FIXING SOME OF WHAT'S WRONG)

When Bohr returned to Copenhagen in 1916, he found papers from Arnold Sommerfeld awaiting him; Sommerfeld was a distinguished professor of theoretical physics at Munich University. More precise measurements had been made that showed a "fine structure" splitting of each line of the hydrogen spectrum into a closely spaced set of lines. And then other splittings were observed when hydrogen atoms were placed in magnetic or electric fields.

Bohr's model couldn't explain these additional lines and seemed to be in trouble. But the more mathematically adept Sommerfeld had solved most of these problems by allowing elliptical as well as circular orbits lying in planes with different orientations and by recognizing that electrons moved fast enough to be significantly affected in their energies by Einstein's special relativity. He used two more quantum numbers, l and m (in addition to n) to describe the split states of closely spaced energies. We'll get a look at these numbers and their physical significance later on.

Fig. 2.10. Niels Bohr in 1922. (Photograph by A. B. Lagrelius and Westphal, courtesy of AIP Emilio Segre Visual Archives, W. F. Meggers Gallery of Nobel Laureates.)

While in England Bohr had been appointed to the new post of professor of theoretical physics in Copenhagen, and with the success of the Bohr-Sommerfeld theory his stature in the field had grown. In 1917, with the help of funding from friends for the land and the buildings, he was able to get approval for his Institute for Theoretical Physics, known as the Bohr Institute, with construction to be completed in 1921. There he hoped to replicate the climate for learning and exploration that he had enjoyed under Rutherford in Manchester. (Fig. 2.10 shows Bohr at about that time.)

Because of the war, German scientists were excluded from international meetings. But Bohr had no particular bias. He invited Sommerfeld to visit Copenhagen, and following that visit was invited to visit Berlin in April 1920. There he met Einstein and Planck for the first time. He stayed at the latter's home. The days were filled with the discussion of physics. Bohr and Einstein would walk the streets of Berlin together or dine at Einstein's home. Einstein subsequently stopped by to visit with Bohr on the way back from a trip to Norway in August.

EXPLAINING THE PERIODIC TABLE (NOT QUITE)

With further work, Bohr would use the Bohr-Sommerfeld theory to begin to qualitatively explain the periodic table and the properties of the elements, in particular how apparently inert, nonreacting elements periodically occur just after and just before highly reactive elements as one considered successively heavier and heavier atoms. The inability to explain the properties of the elements, their arrangement according to these properties in a periodic table, and these particular aspects of the table had long been a failing of classical physics. (As background mainly for Chapters 13 and 14, I provide in Appendix B a brief description of the table, the history of its construction, and a quick look at the charismatic character most responsible for its development.) Bohr was invited to Göttingen to deliver in June 1922 a celebrated set of seven lectures on the subject. Einstein would not attend, out of fear for his safety, but when he heard of Bohr's ideas, he would comment that they appeared "as a miracle" of explanation.

In October 1922, Niels Bohr was awarded the Nobel Prize in Physics *"for his services in the investigation of the structure of atoms and of the radiation emanating from them."* Based on the Bohr-Sommerfeld theory, he had been able to predict the existence of the yet-undiscovered element hafnium, atomic number 72 (that is, having 72 electrons and an equal number of protons), and he made that prediction, which turned out to be correct, during his Nobel Prize lecture in December.

Still, there remained two major problems with the model. These were eventually solved by a young Wolfgang Pauli, through the inclusion of two additional assumptions beyond those already made by Bohr and Sommerfeld.

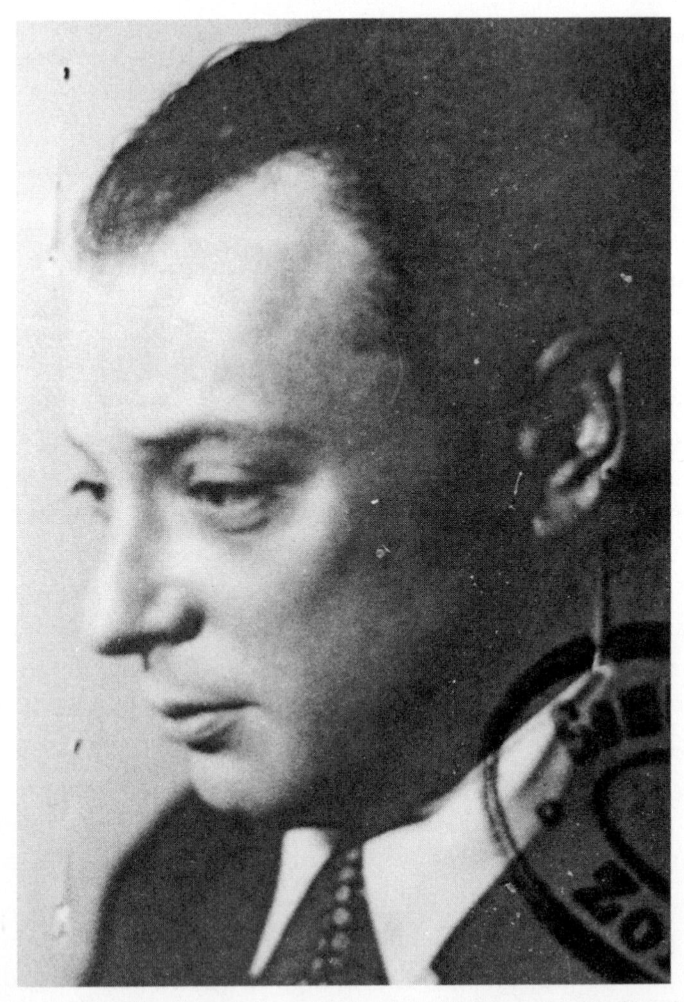

Fig. 2.11. Wolfgang Pauli. (Image from AIP Emilio Segre Visual Archives.)

The Viennese Pauli, fourth from the right in the back row of Figure 1.1, had a brilliance compared sometimes to that of Einstein. (Fig. 2.11 shows an older, more middle-aged Pauli.)

Pauli was born on April 25, 1900. His father had been a physician but shifted to science, at the same time changing his name from Pascheles to Pauli and converting to Catholicism to avoid a rising tide of anti-Semitism. Pauli grew up knowing nothing of his Jewish ancestry. His mother, a pacifist and socialist, was a well-known journalist and writer; and he and his sister, who was six years younger, were exposed to the frequent visits of leading persons in the arts, medicine, and the sciences.

Pauli became interested in physics at an early age, under the influence of his godfather, the renowned Austrian physicist Ernst Mach. Out of boredom with school, he was provided tutors; and out of boredom with his tutoring, he read Einstein's papers on relativity. In January 1919, then just eighteen years old, he published a paper on relativity that brought him recognition in the field. He left Vienna that year (for lack of qualified teachers) to begin work toward a doctorate under the much-respected Arnold Sommerfeld in Munich. He was more inclined toward the nightlife in Munich than formal study, though. Sommerfeld had set him the task of applying the new quantum physics to the ionized hydrogen molecule. That Pauli could not explain experimental findings on the ion was taken as evidence of deficiencies in the Bohr-Sommerfeld model. With doctorate in hand, Pauli in 1921 moved on to study as an assistant to Professor Max Born in Göttingen.

Born is the person second from the right (next to Bohr) in the second row of Figure 1.1. He had sought out Pauli, intent on creating an institute of theoretical physics in Göttingen to rival that of Sommerfeld in Munich. Earlier, Born had formed a strong friendship as a young professor with Einstein in Berlin. Born and Einstein shared a passion for music as well as physics, and even when Born was called into the service (stationed near Berlin), he would often be invited for evenings and music at Einstein's home. Born's approach to physics led from his proficiency with mathematics; Pauli's, more from an intuitive sense of the physics itself.

SPIN (MAKING IT WORK #1)

By 1922, the electrons in the Bohr-Sommerfeld atom were considered to occupy groups of states, with the states within each group only slightly different in energy. Each group, called a "shell," had energies near one of the original Bohr orbital energies. Bohr had begun to picture each successive element as resulting from the addition of one more proton to the atomic nucleus of its predecessor and one more electron into the surrounding shells. The chemical properties of the elements would result from the closeness of each element to either just filling a shell with electrons or going beyond that to add electrons toward the next shell. In particular, elements that tended not to react and combine chemically—helium, neon, argon, and the like—the noble gases, would be marked by the just filling of an electron shell.

The first problem with this was that the electrons seemed to produce the noble gas elements only when twice as many electrons were present than was needed to fill the states in the shells. The theory was off by a factor of two. Pauli postulated in the spring of 1925 (without any reason or proof except that it made Bohr's theory work) that for every one of the Bohr-Sommerfeld states there must be another property, which could take on either of two values marked by a fourth quantum number. Each Bohr-Sommerfeld orbit could take on either of these two values, and so altogether there would be twice as many total spin and orbit states for the electron, just what was needed to explain the noble gas elements.

Two Dutch doctoral students at the university in Leiden, George Uhlenbeck and Samuel Gaudsmit, suggested that this fourth property would have to be intrinsic within the electron, not part of the electron's orbit. They called it *spin*, though there was nothing to indicate that the electron would in fact be spinning. This property would be quantum in nature with no analogue in classical physics. The two students published their results in the summer of 1925. So now there were four quantum "numbers" for each state, three labeled generally by the letters n, l, and m, and the fourth labeled conveniently here simply as "spin."

Uhlenbeck and Gaudsmit were considered for the Nobel Prize, but the idea of spin had been independently proposed earlier by Ralph Kronig, a graduate student from Columbia University who was working on a postdoctoral tour in Europe. Amid the controversy, the committee decided to award no prize at all for this contribution.

EXCLUSION (MAKING IT WORK #2)

A second major problem with the argument of the filling of shells was that all electrons might be expected to occupy just one state, the state of lowest energy, as is the tendency in nature. Because all electrons would be in the same state, shells would not be filled, and the properties of all of the elements (that is, of every type of atom) would be totally unexplained. Every atom would have all of its electrons occupy just one state, the lowest-energy, ground state, and so all would be equally far from filling all of the sites in their lowest-energy shell.

To make the theory function properly, Pauli, in 1925, postulated "*exclusion*"—that no two electrons would be able to occupy the same overall state. That is, no two electrons would have the same set of quantum numbers *n*, *l*, *m* and spin. (Again, Pauli did this without proof or physical reason, except that it fit the arrangement of the elements in the periodic table.) The electrons in an atom would populate the available states, one electron per state, starting with the lowest-energy states. Once every state in a shell had one electron in it, the next electron would go on to start the occupation of the lowest-energy state in the next, higher-energy, shell. The last, highest-energy, occupied electron state in any atom would register the degree to which that atom's outermost shell was filled. The chemical properties of that atom, that element, would be in large measure determined by the closeness to which the electrons came to completing the full occupancy of a shell: for instance, two states away, one state away, full occupancy (a noble gas), one more than full occupancy, two more than full occupancy, and so on. How the chemical properties of the elements are related to these occupancies is described in Part Four. The key point here is that it worked. With these postulates of spin and exclusion, the Bohr theory would qualitatively explain the elements and their properties. And Pauli would receive the Nobel Prize in Physics for 1945 *"for the discovery of the Exclusion Principle, also called the Pauli Principle."*

Despite its successes in describing the elements and their properties, many physicists were uneasy that the Bohr-Sommerfeld model of the atom was built largely on postulate and assumption, and that it still ignored the classical requirement of radiation and collapse for an orbiting (therefore accelerating) electron. Others, Pauli among them, were more than uneasy: they were totally frustrated with the cobbled-together nature of the theory. They cited the need for an entirely new physics.

Chapter 3

HEISENBERG, DIRAC, SCHRÖDINGER—QUANTUM MECHANICS AND THE QUANTUM ATOM

A rapidly occurring string of developments, starting in 1922, would reveal a much different physical world than Bohr, Sommerfeld, Pauli, Einstein, or anyone else had ever imagined. In this key Chapter 3, I describe those developments and provide a first glimpse into that world, our world.

COMPTON SCATTERING (PROOF OF THE PARTICLE NATURE OF THE QUANTUM OF LIGHT)

The outstanding twenty-seven-year-old American experimental physicist Arthur Holly Compton (shown directly behind Einstein in the second row of Fig. 1.1) had been appointed professor and head of physics at Washington University in St. Louis, Missouri, in 1920. Later, at the University of Wisconsin over the winter of 1922–1923, he found an extraordinary result while working with x-rays (high-energy electromagnetic radiation with wavelengths approximately the sizes of atoms). Monochromatic x-rays shining on graphite reflected back with changed and lengthened wavelengths. It was as if violet light reflecting off of a substance would change its color to red in the process. What was observed is now called the *Compton effect.*

Nothing like that had ever been seen to happen with waves of any kind. In fact, what was seen couldn't be explained by waves at all. The only viable explanation was that electromagnetic radiation, as Einstein had shown theoretically in 1916, had momentum in the same sense that a particle has momentum. The electrons in the atoms of graphite (carbon) would recoil from impact and thereby absorb some of the momentum of an incoming x-ray quantum, so that the latter would reflect away in another direction with reduced energy and momentum and a correspondingly increased wavelength.

But Bohr and two colleagues questioned Compton's conclusions. Bohr simply would not accept that light itself would be quantized and have momentum. Compton had assumed conservation of energy in the collision, and Bohr argued (defying one of the universal fundamentals of physics) that conservation of energy had never been proved on an atomic level.

Compton then found a way to confirm the existence of the recoiling electrons and prove that energy and momentum were indeed conserved. And Hans Geiger and Walther Bothe in Germany obtained the same results. The collection of results from the experiments was irrefutable evidence of the particlelike nature of the electromagnetic quantum, by inference also including the light quantum. With these experiments and their interpretation, the quantum of light was finally accepted by the physics community as behaving as both a particle and a wave (as Einstein first proposed it with his analysis of the photoelectric effect in 1905). The "particle" part of wave-particle duality had been confirmed. Einstein had finally been vindicated in the view that he had advocated nearly alone for over twenty years. Compton would in 1927 receive the Nobel Prize in Physics "*for his discovery of the effect named after him.*"

In a paper presented in 1928, Compton used the Greek word for "*light*"—*photon*—to refer to single particles of light or x-rays, and this has become accepted use for particles of electromagnetic radiation in general. (But Gilbert Lewis is described as naming the photon in 1926.[1]) I use the term "light" in this broader context in the rest of this book, though it is more commonly thought of as applying to light in the visible range of the electromagnetic spectrum (refer to Appendix A).

PARTICLES AS WAVES (TURN-AROUND IS FAIR PLAY?)

Louis de Broglie (pronounced "dee-BROY"; third from the right in the second row of Fig. 1.1 and, more closely, in Fig. 3.1) was a graduate student at the Sorbonne in Paris in 1923. He reasoned that if light could have particlelike properties, perhaps particles, like the electron, could have wave-like properties. He backed his thoughts with calculations. It was a radical idea, but when the outside thesis examiner, Paul Langevin, wrote to ask a separate opinion, Einstein examined the work and replied: "He has lifted a corner of the great veil."[2]

Fig. 3.1. Louis de Broglie. (Image from Deutscher Verlag, courtesy of AIP Emilio Segre Visual Archives, Brittle Books Collection.)

Louis Victor Pierre Raymond de Broglie was born in 1892 into an aristocratic French family with the titles of both a French duke and a German prince. His brother, Maurice, seventeen years older, had become head of the family after their father died. Louis was just fourteen at the time. Maurice had started a career in the military, as was the family tradition, but there he got involved with wireless communications, became interested in science, and eventually left the service to obtain a doctorate under Langevin at the College de France. He would go on to build a laboratory in his mansion in Paris and become a recognized researcher on x-rays.

Louis was tutored at home, and guided by Maurice into general studies. But then he opted into the sciences through exposure to Maurice and the lab. One year of compulsory military service was extended a further five years with the outbreak of the war, after which he worked with his brother and wrote several papers.

Louis pictured the electrons in Bohr's orbits as being related to standing waves, like the harmonic vibrations in a guitar string shown in Figure 3.2(a). (These are "standing" because they are trapped and don't travel anywhere.) The half-wavelength string shown in the bottom figure would vibrate up and down at the string's "fundamental" frequency, determining the string's basic pitch. The extent of the "down" swing in this case is shown as a solid line, and the extent of the "up" swing is shown as a dashed line. These extremes mark the "amplitude" of the vibration. The three "harmonics," shown above the fundamental, contribute to the "timbre" of the string's sound. These harmonics vibrate at successive integral multiples of the fundamental frequency.

De Broglie surmised that the full wavelength of a wave, like that shown second from the bottom in Figure 3.2(a), must fit an exact number of times around the circumference of an orbit. Five such wavelengths are shown to fit around the $n = 5$ Bohr orbit, as shown in Figure 3.2(b). This condition would result in the same quantized energies and orbits that had been determined originally by Bohr. But because de Broglie's electron was viewed as a standing wave and not a particle in orbit, there would be no acceleration of charge and no associated radiative slowdown and collapse. So, with de Broglie's standing wave the conflict with classical physics and the continuing fundamental objection to the Bohr model would be avoided.

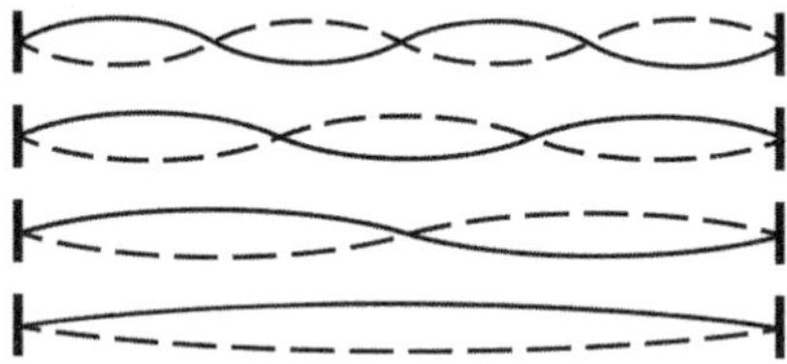

(a) Standing wave harmonic vibrations in a guitar string.

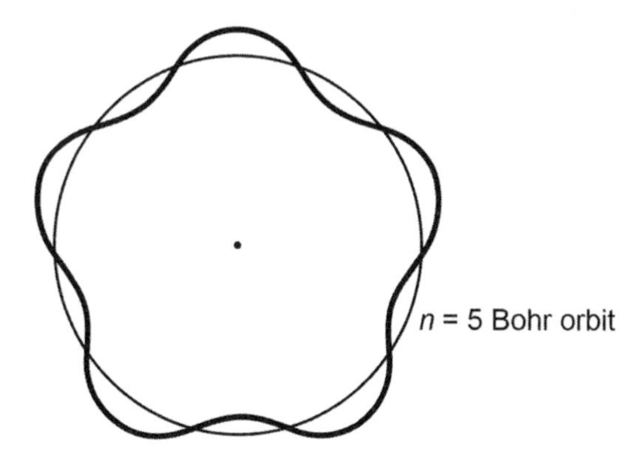

(b) De Broglie electron as a closed vibrating standing wave.

Fig. 3.2. (a) Note the "fundamental" *(bottom)* and three lowest harmonic vibrations of a single guitar string. Each vibrational mode is called a "standing wave" because the string just vibrates in each case between the extremes of the solid line and the dashed line and the waves shown don't travel anywhere. (b) De Broglie envisioned the states of the electron in the hydrogen atom as a set of standing waves that circle within the Bohr orbits and close on themselves. The $n = 5$ standing wave is drawn over an $n = 5$ Bohr orbit.

De Broglie also suggested that electrons by themselves as free particles should undergo a wavelike diffraction and interference, much as had been observed for particles of light, as illustrated in Figure 2.2 by analogy to what happens with waves in water.

This wavelike nature of the electron was glimpsed in 1926 by a team of two American physicists, Clinton Davisson and Lester Germer, when they saw preliminary evidence of diffraction and interference as they directed a beam of electrons at a nickel crystal. (The regular lattice of atoms in the crystal would act to diffract the electrons to interfere in somewhat the same way that the two openings in the breakwater acted to diffract water waves to interfere, as illustrated in Fig. 2.2.) The two researchers were unaware at the time of de Broglie's ideas. Diffraction (when waves spread out past openings) and the resulting interference (when they come together) were confirmed the following year by these men and independently by the British physicist George Thomson (son of J. J. Thomson), who in a somewhat different experiment got similar results as he passed beams of electrons through thin metal films.

Diffraction and interference do not occur unless something is wavelike! The classical concept of particles always behaving as pointlike specifically located objects had been successfully challenged. Their (sometimes, at least) wavelike nature had been demonstrated as fact!

(De Broglie would in 1929 receive the Nobel Prize in Physics "*for his discovery of the wave nature of electrons.*" Davisson and Thomson would share the prize in 1937 "*for their experimental discovery of the diffraction of electrons by crystals.*")

I interrupt my chronological narrative here to describe more recent experiments that further elucidate the wave nature of electrons and light quanta.

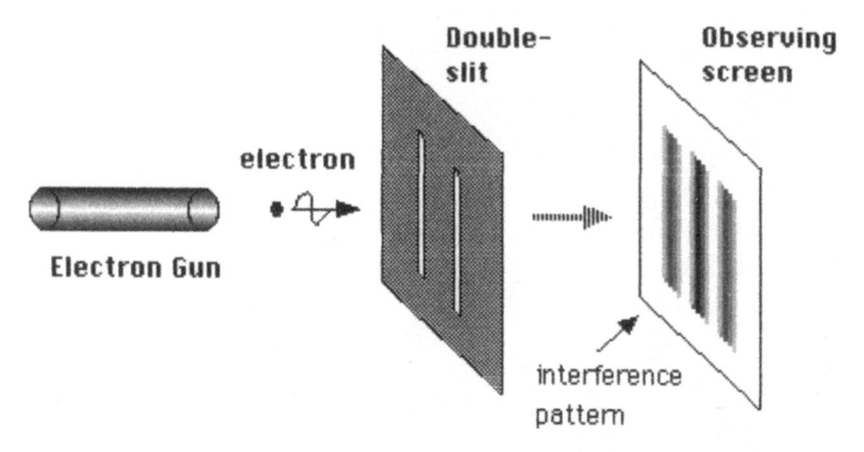

Fig. 3.3. Electrons are shot one at a time at a barrier with two slits in it. Each electron produces a dark dot on a white screen. With many electrons this results is an interference pattern with a dark bar of dots at the center, fading to two lighter bars of dots on either side and still lighter bars (not shown) beyond those. Counting the density of the dots would produce a graph of intensity resembling the wave heights shown in Figure 2.2(c). (Image from *Wikipedia* Creative Commons; file: Double-slit.png; assumed author: NekoJaNekoJa~commonswiki. Licensed under CC BY-SA 3.0.)

The dual particle/wave nature of electrons has been beautifully demonstrated in a two-slit diffraction and interference experiment. The setup for the experiment is shown schematically in Figure 3.3. Everything shown is enclosed in an evacuated chamber. Electrons are shot *one at a time* from an "electron gun" at the midpoint between two slits in an otherwise-blocking sheet of material, and their impacts are recorded as a tiny spots on an observing screen. The accumulation of tens of thousands of spots, each from one of tens of thousands of electrons, produces an interference pattern (shown schematically on the screen) that resembles the interference pattern of a wave.

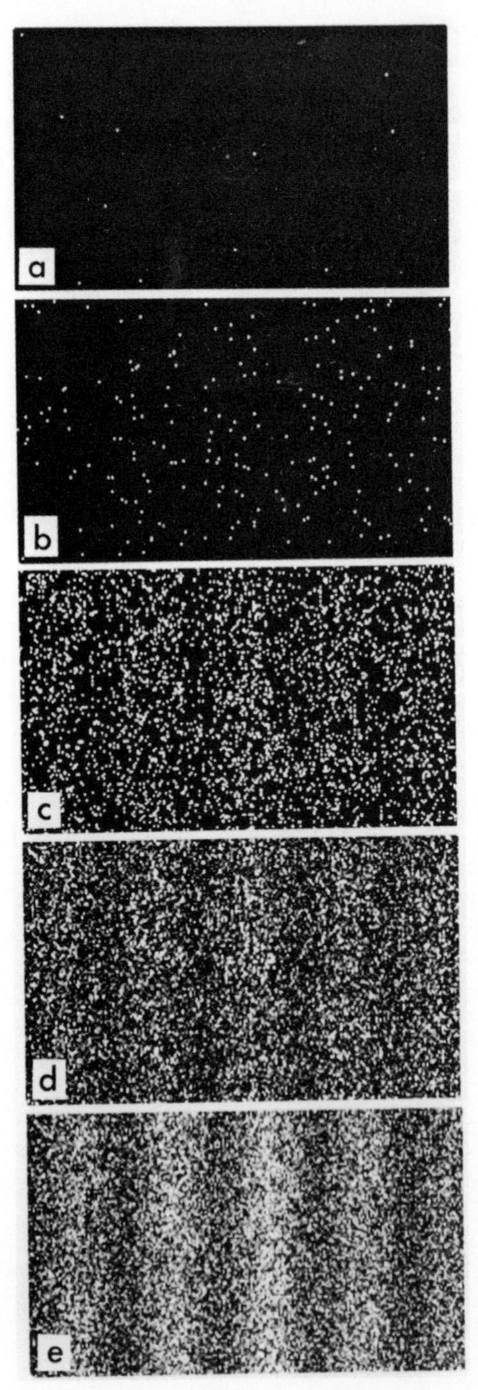

Fig. 3.4. Here we have the same two-slit experiment as in Figure 3.3, except that there is a dark screen that produces a white dot when the electron strikes, and we see the dots of the individual electrons. We see the growth of the interference pattern: (a) with 11 electrons having been shot, (b) with 200, (c) with 6,000, (d) with 40,000, and (e) with 140,000. Note in (a), but it applies to (a) through (e): each electron, shot as a particle by itself, interferes with itself to impact the screen as if it has traveled through both slits as a wave. (Image from *Wikipedia* Creative Commons; file: Double slit experiment results Tanamura 2.jpg; user: Belsazar, with permission of Dr. Akira Tanamura. Licensed under CC BY-SA 3.0.)

Figure 3.4 shows the actual gradual accumulation of spots for the experiment where the screen this time is black and the spots where the electrons impact the screen are white: (a) shows the accumulation of spots from just 11 electrons, (b) for 200 electrons, (c) for 6,000 electrons, (d) for 40,000 electrons, and (e) for 140,000 electrons. The particle nature of the electrons is in evidence by the fact that they are shot from the gun one particle at a time and hit the screen one particle at a time, in each case creating only one spot. But somehow that single electron has a wave property that allows it, if not traveling through both slits, to at least sense the locations of both slits, so that its single spot of impact is located in such a way as to contribute, with the spots of other separately shot electrons, to the formation of an interference pattern.

For the electron, which we think of as a particle, we thus have wavelike behavior. Could the electron actually have split somehow and traveled through two slits to interfere with itself?

Now, remember the two-slit experiment conducted by Thomas Young to show that light was wavelike, as described following the discussion of Figure 2.2. In a recent experiment by Lymon Page of Princeton University, monochromatic light (light all of a single, distinct wavelength) was reduced in intensity so that single quanta, essentially single photons, were shot one at a time through the two slits in a setup like that shown in Figure 3.3. A similar accumulation of single spots occurred, producing a wavelike interference pattern. We pose the similar question: How does a single, indivisible photon travel through both slits and interfere with itself?

Classical ideas, classical physics, can in no way explain the wave nature of individual particles. The double-slit experiment, either for single electrons or for single light quanta, remains one of the best demonstrations that we live in a quantum world. And, as you can see, our quantum world is strange. Quantum mechanics provides an explanation for this strangeness. Stay tuned.

(I would say that we have shed some light on this subject but have not as yet fully illuminated it [puns intended].)

There is a practical side to the wavelike behavior of particles.

It had been determined some considerable time before 1926 that the magnification that could be achieved in optical microscopes would be limited by the wavelength of light. Wavelength needed to be short compared to the size of the object being viewed. With light one could see blood cells, about ten microns (10^{-5} meters) in size, but not much that was smaller.

In 1931, realizing that electrons could be produced with wavelengths 100,000 times shorter than for visible light, Max Knoll and Ernst Rusk invented the electron microscope. Now one could "see" almost to the very size of the atom. The manufacture of commercial units was begun in England in 1935.

(I resume now my chronological narrative of early discoveries in the development of quantum mechanics.) Though Bohr and Sommerfeld in 1916 had invoked a quantized set of orbits in their model of the atom, their model still used classical physics, point particles, and orbits otherwise resembling those of the planets—still "solid ground" for most physicists at the time. However, with the still persistent issue of expected collapse of orbits due to radiation, with the suggestion of the de Broglie atom, and with the experimental demonstrations of the Compton effect in 1923 and electron diffraction and interference in nickel crystals and metal films in 1926, that ground began to shake and crack apart.

QUANTUM MECHANICS (FINALLY, A SOUND THEORY)

Even before the above-mentioned diffraction experiments, three scientists were exploring separate theoretical approaches toward better explaining the nature and the workings of the atom: Werner Heisenberg in Germany (standing third from the right in the back row in Fig. 1.1), Erwin Schrödinger from Austria (pronounced in English as "*SCHROW-dinger*"; standing sixth from the right in the center of the back row), and the reclusive Paul Dirac from England (pronounced "deer-ACK"; in front of Schrödinger and behind Einstein). Remarkably, all three approaches produced the same results. However, of the three, Schrödinger's "wave mechanics" version lends itself best to visualization, and we concentrate on his approach to describe the fascinating key concepts of the atom and eventually the broader workings of our universe (or universes, as you will see). First, a quick glimpse of the three approaches and the personalities involved in their development.

Matrix Mechanics

Working in Göttingen in June 1925, Werner Heisenberg struggled to make sense of the spectral lines from hydrogen, a problem that his professor, Max Born, had assigned him two years earlier. Heisenberg came up with the idea that the atom could be treated as a bunch of quantum oscillators (an approach somewhat similar to that used by Planck on light radiation twenty-five years earlier and by Einstein on the specific heat of solids nine years after that). Heisenberg developed an array describing the possible transitions within the atom, a theory of sorts. With encouragement from Wolfgang Pauli (who had come to be respected and feared as a no-holds-barred judge and critic of work in the field), Heisenberg published his theory, the first paper on quantum mechanics. Einstein and Bohr were skeptical but hopeful that it would open up something new. But the theory, though stimulated by consideration of the atom, was rather obscure and had yet to be successfully applied.

Fig. 3.5. Werner Heisenberg in 1927. (Image from AIP Emilio Segre Visual Archives, Segre Collection.)

Werner Karl Heisenberg (shown more closely in Fig. 3.5) was born in Würzburg, Germany on December 5, 1901, the younger of two boys.

His father became professor of Byzantine philology at Munich University and moved the family there when Werner was eight. The aftermath of the war was particularly difficult in Bavaria, where radical socialists sought to declare a Soviet Republic. Werner and his friends formed one of the many militarylike groups that were organized to oppose the movement.

Through it all, Heisenberg excelled in school and won a prestigious scholarship to the university. His coming of age in physics involved study and work at the three major centers of thinking on quantum physics. He began in 1920 as an undergraduate alongside Pauli in Sommerfeld's institute in Munich. Both recognized Heisenberg's potential, and Pauli and Heisenberg were afterward to maintain a close professional communication. When Sommerfeld left for a stint in America, he arranged for Heisenberg to study with Born in Göttingen. Heisenberg subsequently returned to Munich, completed his doctorate, and then responded to Bohr's invitation that he spend a year in Copenhagen, this following Heisenberg's asking penetrating questions of Bohr after one of his Göttingen lectures in 1922.

Pauli had written to Bohr of Heisenberg, and the young physicist was received warmly and personally. The men developed a professional and personal relationship. (This has been portrayed as somewhat of a father/son relationship in the play and movie *Copenhagen*. It describes Heisenberg's visit with Bohr after Heisenberg had taken over leadership of the Nazi atomic weapons program during World War II.) Bohr, as Heisenberg would soon learn, even more than others, was concerned with the deficiencies being revealed of his own model for the atom.

Born recognized that working with Heisenberg's array of oscillators would require the mathematics of matrices, a set of mathematical tools not generally known to physicists at the time. Born enlisted his student, twenty-two-year-old Pascual Jordan, who had transferred into physics from mathematics, to devise a good theoretical framework. They submitted their findings in collaboration with Heisenberg for publication in October 1925. Their approach would be labeled more specifically *matrix mechanics*.

Operating with matrix mechanics was cumbersome, but Pauli, working in parallel, mastered it and applied it to successfully derive the hydrogen spectrum. Matrix mechanics got results! But what did the atom look like? The theory wouldn't say.

Dirac's Quantum Mechanics

Paul Dirac's paper recognizing the fundamental physics underlying Heisenberg's original theory was received by the Proceedings of the Royal Society in London before the "three-man paper" on matrix mechanics described above had been submitted for publication.

Fig. 3.6. Paul Dirac at middle age. (Photograph by A. Bortzells Tryckeri, courtesy of AIP Emilio Segre Visual Archives, E. Scott Barr Collection, Weber Collection.)

Paul Adrien Maurice Dirac (shown more closely at middle age in Fig. 3.6) was born in 1902 as the second of three children of an English mother and a Swiss French-speaking father.

His overbearing father, who taught French, insisted that all communication with his son be in French. But Paul found that he could not express himself well in French, and so he chose not to speak very much at all. This carried over as a general characteristic. He tended to be withdrawn. But he was also brilliant.

Dirac was interested in science, but his father steered him into engineering. He graduated in three years from the University of Bristol as an electrical engineer, was unable to find a job in the aftermath of the war, and continued there with the offer of free tuition to earn a degree in mathematics. Then, with the help of a government grant, he was admitted to Cambridge, which earlier he had simply been unable to afford. Even while an engineering student he had read and understood Einstein's relativity, and he'd listened to Bohr lecture at Cambridge; he was impressed by the man, but not by his theory.

Working alone at Cambridge, but exposed to Heisenberg's earlier work by his advisor Ralph Fowler (Rutherford's son-in-law, second from the right in the back row of Fig. 1.1), Dirac by the end of November 1925 had turned out four papers that together constituted his PhD thesis and earned him his doctorate. It was a more complete and careful work than the three-man paper, and it used a different formalism. (Dirac would contribute in an even more fundamental way to quantum mechanics, as will be described in Part Two. He would combine relativity and Schrödinger's equation in a way that naturally produced the spin of the electron. Later he would be elected Lucasian Professor of Mathematics at Cambridge, a post held earlier by Isaac Newton and to be held later by Stephen Hawking.)

Wave Mechanics

Schrödinger, a professor at the ETH in Zurich, was already a published, solid, recognized physicist when in October 1925 he began to address the problems of the Bohr atom.

Fig. 3.7. Erwin Schrödinger in 1933. (Image from *Wikipedia* Creative Commons; file: Erwin Schrödinger (1933).jpg; author: Nobel Foundation.)

Erwin Schrödinger, shown in Figure 3.7, was born on August 12, 1887, the only child in an upper-middle-class family in Vienna, at a time when the waning Hapsburg Empire and first Austrian republic were centers of contemporary culture. After excelling in the equivalent of American high school and college, Schrödinger went on to graduate school and was in 1914 awarded a doctorate in physics from the University of Vienna.

That year marked the beginning of World War I, and Schrödinger was soon called up as an artillery officer on the Italian front, where he served with distinction. By the war's conclusion in 1918, his father's business had closed. Both of his parents died soon thereafter, and prospects were poor for the young and brilliant physicist in postwar occupied Austria.

Eventually, in 1920, at thirty-two he married his longtime sweetheart, Annemarie Bertel, a working twenty-three-year-old country girl from a good family. But this required that he take a series of positions throughout Europe where he could be paid well enough to support her. His work on wave mechanics was at the height of his career in and around Zurich in 1926.

In 1933, Schrödinger left a distinguished professorship as Planck's successor in Berlin for a position at Oxford because he disliked Germany's sanctioned persecution of the Jews. Hitler had just become chancellor. Moore notes: "Schrödinger was exceptional—very few non-Jewish professors refused to knuckle under to the Nazis," punctuating this with the further comment, "In the autumn of 1933, 960 professors published a vow in support of Hitler."[3] Schrödinger was not one of them.

Schrödinger returned to Austria in 1936. But after the annexation of Austria by Germany in 1939, Schrödinger found it difficult to remain there, having earlier been entered into the Nazi records as "politically unreliable." He moved with his wife to take visiting positions at Oxford and at Ghent before being invited by the prime minister of Ireland, Eamon de Valera, to help set up the Institute for Advanced Studies in Dublin. He became director of the School for Theoretical Physics and remained there for seventeen years, during which time he became a naturalized Irish citizen and wrote another fifty publications on various topics, including efforts to develop a unified field theory.

James Watson refers to Schrödinger's book *What Is Life*, written in 1944, as having inspired Watson to discover (with Francis Crick, Maurice Wilkins, and crystallographer Rosalind Franklin) the double-helix structure of DNA, which, as you may know, allows replication of the genetic code during the process of cell division. (Watson, Crick, and Wilkins received the Nobel Prize, but Franklin died before the prize was awarded, and therefore she was not a candidate. Her work was critical to the discovery, however.)

Schrödinger found Heisenberg's 1925 oscillator approach to quantum mechanics to be obscure and lacking any comfortable physical description. He turned to examining de Broglie's standing-wave theory, following a footnote in one of Einstein's papers. He set to work to try to solve for the electron in the atom as standing waves in three dimensions. His resulting description of the atom as a set of standing waves seemed more tangible than matrix mechanics, involved a more widely known and easily used mathematics, and seemed at the time to offer some possibility of a comfortable connection to the classical physics that many (especially Schrödinger himself) were reluctant to abandon.

A Conflict Over Theories?

Conceived over Christmas 1925 and published in March 1926, Schrödinger's wave mechanics was welcomed by Einstein and Bohr and eventually most in the field (but not without questions); it was challenged especially by Heisenberg, who was increasingly becoming frustrated that the physics community was endorsing Schrödinger's concepts rather than his own. During Schrödinger's lecture in Munich in July, Heisenberg asked questions that Schrödinger had difficulty answering. For example, Schrödinger could not explain the photoelectric effect: How could the electron emerge as a "popped loose" wave? And how would a wave carry the charge of the electron? Schrödinger at first thought this could be done with the charge smeared-out in some fashion. But this would seem to deny the electron's observed particle nature and possession always of a single, indivisible localized basic unit of charge.

Following the meeting, Bohr invited Schrödinger for a visit to Copenhagen. In early October, the two of them, together with Heisenberg, who had returned there from Munich, would try over several days to sort out conflicts and questions in both theories. It was more Bohr than Heisenberg who was relentless in questioning the weakness of wave mechanics. But Heisenberg and Bohr had no answer to Schrödinger's call for a mechanism describing "quantum jumps," the transformation of the electron from one energy state to another. Schrödinger's wave mechanics would seem to allow for such a mechanism, but Heisenberg's matrix mechanics would not. A broader, all-encompassing formalism seemed to be needed to answer the questions in both theories.

Transformation Theory

The answer was provided in part independently in late 1926 by Pascual Jordan and Paul Dirac. While working at Bohr's invitation during a six-month stay in Copenhagen, Dirac showed that matrix and wave mechanics were just special mathematical cases of his own broader "transformation theory." This sorted the mathematics, but a good physical interpretation of the theories and resolution of their apparently conflicting physical aspects was still lacking.

(Heisenberg in 1932 would receive the Nobel Prize in Physics "*for the creation of quantum mechanics, the application of which has, inter alia, led to the discovery of allotropic forms of hydrogen.*" He felt that Jordan and Born should have shared the prize. In 1933, Schrödinger and Dirac would together be awarded the prize "*for the discovery of new productive forms of atomic theory.*")

WAVE MECHANICS AND THE HYDROGEN ATOM (HOW IT WAS DONE)

Schrödinger's "wave mechanics" approach to quantum mechanics gives us the best picture of what is going on. I describe it as follows, starting with a simple analogy, and conclude with another analogy that shows his results in comparison with something more familiar to us.

> What Schrödinger did in 1925 is a little bit like what many of us learned back in high-school algebra. Remember, we would use the letter x to represent some unknown sought answer, and then we would set up an equation to describe the circumstances related to x that offered clues to its value. We would then solve the equation for x. Sometimes the solution would produce two values of x. These were the only values for x that would fit the circumstances, the only correct and allowed answers.

Schrödinger defined a complex mathematical function represented by the Greek letter Ψ (pronounced "psi"). Ψ would contain all of the information about a particle's position and movement through space and time. Every one of the physical properties of the particle could be extracted from Ψ by performing certain separate mathematical operations on it. One operation would provide the electron's position, another its velocity, a third its energy, and so on.

Instead of solving for numbers for the unknown number x as we did in high school, Schrödinger would solve for the mathematical form of the unknown *function* Ψ. He set up an equation making the total energy of the atomic system (given by an appropriate mathematical operation on Ψ) equal to the *sum* of the energy of motion of hydrogen's single electron (given by a different operation on Ψ) and the energy associated with the attraction of the electron to the single-proton nucleus (given by yet another operation on Ψ).

It sounds like a lot of mathematical mumbo-jumbo, but when Schrödinger solved his equation the results were a first-order triumph. With the added inclusion only of the assumption of spin, later to be derived by Dirac as an intrinsic property of the electron (and so no longer an assumption), Schrödinger's model almost exactly predicted all of the observed properties of the hydrogen atom, and did so without the ad hoc postulates required for the Bohr-Sommerfeld model!

When Schrödinger solved his equation, he got not one or two but an infinite series of separate standing-wave solutions for Ψ at discrete, separate, energy levels. These corresponded exactly with the energy levels of the Bohr model shown in Figure 2.7(b), and, with spin included, further split those levels into the closely spaced energies derived by Sommerfeld. These solutions for Ψ represent the only states that mathematically satisfy the wavelike physics of the electron in the hydrogen atom. They are thus the only *allowed* states for the electron in hydrogen. And the differences in the energies of these states give exactly the hydrogen spectrum that had been characterized (as described in Chapter 2) by Balmer, along with the further fine structure and magnetic- and electric-field splitting of the lines that were observed later on. But, most importantly, unlike the Bohr atom, there aren't any orbits, and so there is no acceleration of the electron and no reason for radiation and collapse!

PROBABILITY (A WAY TO VISUALIZE THE ATOM)

Just a little later, in August 1926, Max Born was beginning to suggest that particles are not physical waves, rather, that wavelike characteristics are only in the mathematics of wave mechanics (which determines among other things where a particle may be found). According to Born, the magnitude (think of wave amplitude) of one of Schrödinger's mathematical standing-wave solutions at any point in space should be a measure of the probability that a particle would actually be located at that particular point. In this way he would retain the ideas of both waves and particles and resolve the differences between Heisenberg's more particulate theory and Schrödinger's more wavelike theory. Born's probabilistic view would be used to interpret Schrödinger's results and would be endorsed by Bohr to become a part of what would be called the *Copenhagen interpretation* of quantum mechanics (to be further explained later on). Born, in 1954, received the Nobel Prize in Physics "*for his fundamental research in quantum mechanics, especially for his statistical interpretation of the wavefunction.*"

But Born's probabilities are not "statistical" in the classical physics sense of indicating the probable behavior of a collection of particles. Classical statistics, for example, would show that atoms of helium gas released from a popped balloon are unlikely, in bouncing around the room, to find themselves some time later all back in the region that had been occupied by the balloon. Born's probabilities in the case of electron states are statistical only in "not being definite" about location. They are more akin to the probabilities of getting a seven or "snake eyes" in the roll of a pair of dice.

(Note, Schrödinger's equation applied to free particles also results in wavefunctions, and these evolve with time in such a way as to indicate the probable motion of the particle. His equations can, in principle, be applied to whole systems of particles and objects, with resulting wavefunctions that describe the evolution of the entire system, even the evolution of the entire universe.)

A Way to Visualize the Atom

The most common way of describing the spatial form of the quantum atom today comes from Born's probability interpretation of Schrödinger's standing-wave solutions Ψ for hydrogen. In the best representation of the atom that I have seen anywhere, the probability of hydrogen's one electron being at any particular point in space is marked as a spot of certain brightness in an otherwise three-dimensional blackness. If for each solution to Schrödinger's equation for hydrogen we visualize all points in space as having a brightness in proportion to the magnitude of Ψ, we create a fuzzy white probability cloud with inherent symmetries. The cross sections of five of these clouds, representative of five of the lowest energy states in hydrogen, are shown magnified one hundred million times in Figure 3.8.

These states are in fact, mathematically, standing-wave solutions to Schrödinger's equation. They provide rigorously derived three-dimensional representations in furtherance of de Broglie's idea of a standing wave closed on itself (which was illustrated in Fig. 3.2). The lobed spatial states in Figure 3.8 may be thought of as standing waves where the bright areas are where the amplitude of wave fluctuation is highest and the dark areas are places of near zero amplitude.

Each cloud in Figure 3.8 is a separate possible pattern of probability for the electron. With exceptions to be described later, the electron will be found in (i.e., will occupy) only one standing-wave spatial state, represented by only one cloud (or, in two dimensions, by one cloud cross section) at any particular time. The single-proton nucleus, which would be located at the geometrical center of each cloud, is too small to be visible even at the very high magnifications for the clouds shown in Figure 3.8.

For clarity of viewing, I have shown each cloud cross section separately, but all states, each represented by its particular cloud, collectively surround the same nucleus in a single atom, so all states (and their cloudlike representations and the cross sections of them) are actually superimposed one over the next, each symmetrically surrounding the same tiny nucleus of the hydrogen atom. *These probability clouds are our best visual description of the hydrogen atom.*

As you will see, straightforward modifications of this visual model for hydrogen can be used to describe the possible states of the electron in the atoms of the rest of the elements. When physicists set up to solve for the presence of *two or more* electrons in the states of these atoms, with the inclusion of spin for each electron, they found that no two electrons would be allowed to occupy the same total spin and spatial state. They were thus able to *derive* Pauli's exclusion principle! It was no longer an assumption.

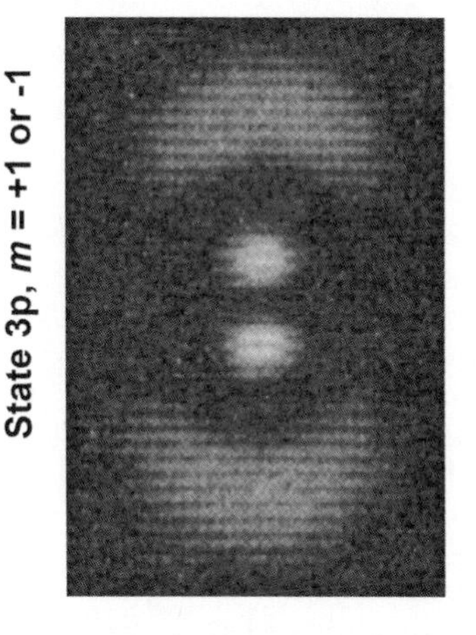

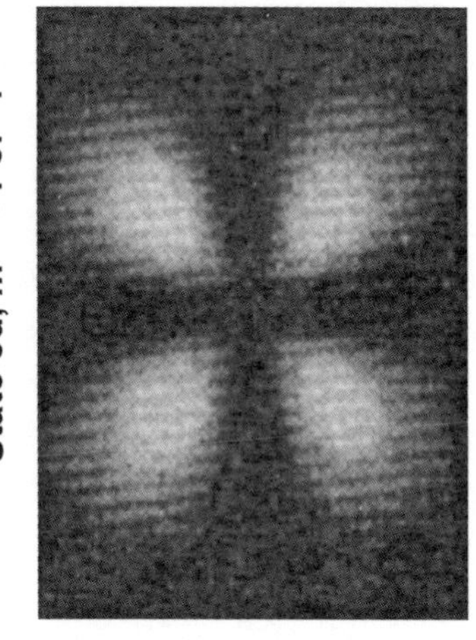

Fig. 3.8. Relative sizes of the spatial states of the hydrogen atom. (Images from Fig. 5-5 of Leighton, Reference F, with permission from Margaret L. Leighton.)

Each cloud cross section in Figure 3.8 is labeled by numbers and letters identifying it as representing that Schrödinger solution with a specific *n*, *l*, and *m* quantum number (spin will be included later). The number at the top indicates *n*, the Bohr energy level; the letter afterward indicates the nature of a spectral line and an angular momentum (to be defined later) corresponding to a quantum number for *l*; and the number for *m* at the bottom indicates a magnetic characteristic (also to be defined later). The properties corresponding to these quantum numbers for each state have all been observed experimentally, directly or indirectly.

The two $m = 0$ clouds at the lower left of the figure look in three dimensions like a fuzzy ball and a fuzzy ball within a hollow fuzzy ball. The three-cloud cross section marked $m = +1$ or $m = -1$ have lobes of probability extending out in different directions. The smallest of the cross sections, the cross section of the "fuzzy ball" shown at the lower left, is for the 1s lowest-energy "ground state." Everything in nature, including the electron in hydrogen, tends toward a lowest-energy state. So, what we see to the lower left in Figure 3.8 is the most likely size and shape of the probability cloud for the hydrogen atom (of course magnified 100 million times). The atoms of the rest of the elements have similar, solved-for ground states of minimum size.

By contrast, in classical physics there is no reason to expect any minimum size for the atom. And, as noted earlier, the radiation of energy from an accelerating electron in classical orbit would collapse the atom down to the size of the nucleus, as much as 100,000 times smaller than the size that we observe for atoms. Quantum mechanics explains why this collapse does not actually take place: (as noted earlier) there simply is no solution to Schrödinger's equation that allows a state smaller than the ground state. And, as you will see in Part Four and Appendix D, the minimum and maximum sizes of the entire complex of many states occupied by the many electrons in the atoms of the other elements can be roughly predicted by extension from Schrödinger's solutions for the hydrogen atom. These predictions explain why we (ourselves composed of atoms), like everything around us, are the size that we are and not 100,000 times smaller. There simply are no smaller states and atomic sizes for the electron: they don't exist.

Finally, note that Schrödinger's solutions have come to be called "orbitals" because of their correspondence with the orbits of the Bohr-Sommerfeld model of the atom. But, as noted above, these standing-wave solutions for the electron in any of these solutions is no way in the form of orbits. Nor do these probability clouds rotate. There is nothing orbital about them. Thus, so as not to mislead, in this book I refer to Schrödinger's solutions simply as "spatial states." The probability clouds are "representations" of these spatial states.

The probability clouds are "fuzzy," spread out and diffuse. Although for any state the electron is most likely to be found in the bright regions of high probability, it has some small probability of being found even in most of the darker regions, with the probability eventually fading off to zero at infinity in any direction. The location of the electron is thus *uncertain*: though the electron is most likely found where the probability clouds are brightest, it actually could be almost anywhere. The distance over which the probabilities get to be very small is very, very small for large objects compared to their size. We know their location with relative certainty. But for the electron in the atom, this distance and the uncertainty in the electron's position are as large as the atom itself. (More on uncertainty later on.)

Analogy to the Vibrations in a Drum

What occurs in Schrödinger's atom is in some ways analogous to what happens in musical instruments, and specifically in the operation of the drum. The ground state of the hydrogen atom and its infinite number of higher energy states mathematically resemble the fundamental mode of vibration of the drum head and all of its infinite number of higher-level harmonic vibrations.

Like the discrete set of patterns of vibration in a drum head, Schrödinger's equation applied to the natural tuning inherent in the physics of the atom results in fundamental and additional standing-wave solutions for the electron. But the standing waves in this case are Schrödinger's spatial-state solutions. They are three-dimensional, and they have nothing to do with physical vibrations. Instead they provide information about the probability of location of the electron around the nucleus of the atom, and about other properties that this electron may have.

By their superposition around the nucleus, there is analogy of electron states to the superimposed vibration states of the drum head. But in the drum head all vibration states can be to some degree simultaneously active, whereas the electron, with only a couple of exceptions to be described later, *occupies* and has the properties of only one state at any given time. While the patterns of the drum head are confined to exist only within the drum and with zero amplitude of vibration at the rim, the probability patterns of the electron in the atom are constrained only to have zero value at infinity in any direction. Like each pattern of harmonic vibration in the drum head, each Schrödinger wavefunction solution fluctuates. But it does so in a complex way, and the amplitude of these fluctuations at any location is a measure of the probability that the electron can be found at that location. Said another way, the brightness in the probability clouds for the possible states of the electron is always a positive measure, in each case the absolute positive magnitude of one of the wavefunction solutions for Ψ.

Schrödinger's spatial-state solutions for the hydrogen atom collectively provide most of the information needed to predict the physical and chemical properties of hydrogen. Each spatial-state solution has its own set of properties. These states and their associated energies not only describe the hydrogen atom. They are the key to visualizing and understanding the makeup and properties of the atoms of the rest of the elements, and the atoms of the elements are the building blocks of our bodies and all of the substances of the universe around us.

As will be explained in Part Four and Appendix D, the atoms of the other elements are "tuned" differently by having more protons in the nucleus and an equally greater number of electrons around it. The greater number of protons exerts a greater attraction on each electron and pulls down the energies and sizes of the states for each electron. And the energy levels and patterns of the probability-cloud representations of the various states also differ from those of hydrogen because of the manner in which the electrons repel and interact with each other. Ultimately, it is the numbers of states at each energy level and the manner in which the states are populated with electrons that determines the properties of the elements and all of chemistry.

At this point I want to give you a preview to the realization that quantum mechanics is not just "all theory." So I interrupt our historical narrative to next, in Chapter 4, explain the operation one of the most extensively used of quantum devices, the laser. Its operation is made possible by the quantum nature of our world, by the separate and distinct quantized energy levels of the electron in the atom. Through the work of scientists and engineers we have learned how to make use of this quantum arrangement, not just for the laser, but also for the many inventions and developments to be described in Part Five.

Chapter 4

APPLICATION— SIX HUNDRED MILLION WATTS!

Power is a measure of how quickly energy can be generated, used, or delivered from one object to another or from one place to another.

Imagine that we could trigger the nearly instantaneous (within one hundredth of a millionth of a second), simultaneous transition of electrons from one state to another in twenty billion billion atoms. (This is about 0.3 percent of the atoms that can be packed into a volume the size of a sugar cube.) And suppose that the photons released in these transitions each had the energy of photons emitted in the red spectral-line transition of an electron in the hydrogen atom, from a state having energy level $n = 3$ to a state having energy level $n = 2$, as described in Chapters 2 and 3 with reference to Figures 2.7 and 3.8. Then the power released would be six hundred million watts! This power compares with about ten thousand watts for the brightest flash lamps producing ordinary light. (Flash lamps are used, for example, to produce a sudden bright light for photography. They typically involve a brightly burning powder or metal, sometimes enclosed as in "flash bulbs.")

Suppose, further, that these photons all traveled in the same direction and, further still, that all of these photons were like waves that are exactly in phase, with their crests and troughs all aligned so that the intensity of the beam is like the added-together wave heights of three billion billion photons all synchronized together. Then we would have the *coherent* light of a high-powered *laser beam*.

In a moment I'll explain how the laser works and indicate some of its applications. But first, as follows, I want to explain what is meant by "power" and "coherence."

To start, I have a confession to make: I used "Six Hundred Million Watts" as part of the title for this chapter because it sounds like a lot of power, and it is. However, as noted above, power is just a measure of how fast energy is delivered. This very high wattage in small early lasers resulted mainly from the very short time in which the energy of the laser was released (in about one hundredth of a millionth of a second, as noted above). So this laser had a high rate of energy delivery. The same laser energy, if it were to be released over a much longer time period, say in one second, would be delivered at a rate of only six watts.

This is not to say that there aren't powerful, high-energy lasers. One such laser, resupplied with energy, has enough essentially continuous power to cut apart inch-thick plates of steel. Other even more powerful lasers are being developed for defense and commercial electric power applications, as described in Chapter 20.

Coherence in the context of the laser is analogous to regimentation. Consider, say, one thousand musicians in a marching band, all in neat, even rows. The entire band can perform intricate maneuvers in unison. Or parts of it can be split off to come back and cross or otherwise interact while still marching in sync. Or consider one thousand soldiers all marching in lockstep. Every left or right foot hits the ground in unison: a one-thousand-fold increased impact with each step. (Which is why soldiers are told to break cadence in crossing a bridge.) That's coherence applied to people.

Now examine Figure A1(c) of Appendix A. Every photon in a coherent light beam would have the same wavelength, *w*; move with the same velocity, *v*;* and have its peaks and troughs lined up to add exactly in sync with those of every other photon. That's the coherent nature of light released from a laser. And, because of that coherence, the light from the laser can undergo intricate maneuvers, retain its focus, and have high impact in the sense of the synchronized steps of the soldiers. (*Remember, velocity is a vector [an arrow] that indicates both direction [in the way that it points] and speed [by its relative length, which indicates its magnitude, usually for light close to $c = 3 \times 10^8$ meters per second, the speed of light in a vacuum].)

Laser is an acronym referring to Light Amplification through Stimulated Emission of Radiation. As described in Chapter 2, in 1916, before quantum mechanics was formulated but during the years when quantum theory first began to be developed, Einstein suggested (based on calculations) that the transition of an electron from one state to another in an atom could be *stimulated* by the presence of photons, each having an energy corresponding to the energy that *would be* released during the transition. Providing that enough electrons could be placed in identical (higher-energy-level) excited states in a collection of atoms, it was then conceivable that the presence of one photon might trigger a transition (from that energy level to a lower energy level) that would produce an identical photon from the energy released. And these two photons would each stimulate two more transitions to produce a total of four photons, and these would produce four more for a total of eight, and so on in a chain of transitions that could rapidly create an enormous number of photons. (Or the presence of one photon could stimulate the transition of many photons, in which case the transitions would occur even more quickly.)

It took over thirty years for the stimulation process to be developed, first with microwave photons (the maser, Microwave Amplification by Stimulated Emission of Radiation) and then, five years later, with visible light (the laser). The first laser was produced in 1960 using a piece of aluminum oxide (a synthetic ruby). It was found that light from a xenon flash tube could be used to excite a large number of electrons into an excited (higher-energy) state in a two-inch-long ruby crystal (xenon, because one of its radiated wavelengths is of exactly the right energy to excite the electrons up into the higher energy state). Photons were then emitted during the transitions of these electrons to a lower energy state, photons not much different in energy than those emitted in the hydrogen transitions mentioned earlier. The ends of the crystal were cut and then polished so that they would be separated by an exact number of wavelengths to reflect the photons back and forth to stimulate more and more and more of these transitions, with more and more photons being created, all with the crests and valleys of their wavelike properties synchronized to produce a *coherent* wave thousands of times stronger than might be produced by the waves if they were superimposed randomly.

Because photons travel at the speed of light (in the visible range of wavelengths, they *are* light), each photon could be reflected up to a dozen times in the hundred-billionth of a second of the laser pulse, and so it would have lots of opportunity to stimulate additional transitions. One end of the ruby was made to be only partially reflecting, so that some of the coherent photons would escape (shine) from that end (to produce a laser beam) while most of the photons were stimulating additional transitions.

COMMERCIAL, MEDICAL, AND SCIENTIFIC APPLICATIONS

In the years since 1960, lasers have been developed using gases, liquids, semiconductors, and various other solids. We have found ways to continuously re-excite the electrons so that they can be essentially continuously transitioning to release a continuous beam of coherent light. Numerous applications have been developed for the laser, including precision alignment in machining and automated manufacturing; the clean, fast cutting of various materials; cataract surgery; shaving of the cornea of the eye to eliminate optical defects; laser surgery to correct retinal detachment; measurement of the shape of the eye to monitor for glaucoma; bar-code readers at supermarket checkouts; modern spectrographs to vaporize minute samples or to *peel off* and analyze materials layer by layer; laser-guided weaponry; holography; a host of precision measurements including the Doppler measurement of fluid flow; the measurement of movement of the plates of the earth in plate tectonics; the pulsed photography of the rapid change of shape of molecules in a liquid, optical scanners to read the information recorded on CDs and DVDs, and so on.

The hologram was widely seen by the movie-going public in *Star Wars*, with R2-D2 projecting a three-dimensional, moving, lifelike image of Princess Leia delivering her message of warning and asking for help. With proper equipment, similar images could be projected in your own living room. In holographic photography, laser light is channeled to both reflect off of the object being photographed and shine directly to interfere with the reflected light while exposing the film. When this film is developed and backlit with the same kind of coherent laser light, the synchronized light reinforces or cancels itself in such a way as to create images in three dimensions.

Before we get on to describing more laser applications and other applications resulting from quantum mechanics, we need to understand the theory a little better and get a better sense of our quantum world and where the theory can be applied. So, next, in Part Two of this book, we return to our historical narrative and, as it happens, the excitement and controversy of the emergence of quantum mechanics.

THIS IS A RECORDING.
DOESN'T BOTHER ME. I'M A HOLOGRAM.
S. Harris

Part Two

INTERPRETATION, AND MIND-BOGGLING IMPLICATIONS (1916–2016)

Chapter 5

THE ESSENTIAL FEATURES OF QUANTUM MECHANICS

As we've discussed, there are several derivations and formulations of quantum mechanics, all of which work to produce the same results—all producing, for example, the same set of states for the electron in the hydrogen atom. All of these approaches accurately describe the observable properties of our quantum world. What differs between them is the physical and philosophical view of the world that they suggest and the functional ease of their use. Even within a single approach there is to this day a range of variously disputed interpretations and their associated physical implications, and those will be revealed as we move on with our chronological narrative.

Despite the various formulations and interpretations, there is a set of essential features that make up quantum mechanics, a working description that has survived the controversy of the theory's arrival. It will be helpful to have this set of features in mind as we continue our chronological narrative, and so I summarize these features briefly as follows. I do so using Schrödinger's formalism, since Schrödinger's wave mechanics with Born's probability interpretation has emerged as the most commonly used working approach for quantum mechanics (with conceptual strategies and rules later laid out by Feynman). We will learn in the next chapter how this approach came to be accepted, and of any deviations from this basic framework.

I first list below the essential features of quantum mechanics in the Schrödinger context. Then I explain what each feature means, often with reference to what we have already learned of the electron in the hydrogen atom. To begin, I briefly review what is meant when we discuss the "state" of an object.

What I present in this chapter is a bit formal, but the chapter isn't long, and knowing these features will be helpful toward keeping track of the prevailing ideas through the historical events that follow.

In the classical approximate description of our quantum world, we define an object's state as the combination of its position, mass, speed, and direction. Depending on its position, how heavy it is, and how fast it is moving, we say that it has a certain energy. We can change the state of the object smoothly by increasing its speed a little bit, or a lot, or by moving it from one place to another through all of the positions nearby to those farther away or in between. We can continuously change its energy by increasing its speed, little by little or even rapidly.

In our actual quantum world, an object's state is defined in much the same way. Schrödinger's equation and its evolving wavefunction solutions describe precisely the movement and properties of objects, whether applied to an isolated particle, isolated pairs or groups of interacting particles, or the entire interacting complex of particles and bodies in our universe. However, when one object is in some way bound to the vicinity of another, there is no continuum of states and energies. In these instances, only particular separate and discrete (meaning not continuously connected) states and energies are possible, as we found, for example, for the bound states of the electron in the hydrogen atom. No states or energies are available in between these particular states and energies. We say that the states and energies are quantized. This runs counter to our classically based intuition, where we expect a seamless continuum of states and energies. (Even the planets orbiting the sun have quantized, discrete energies. But the energy levels are so close together that the change from one energy level to the next would appear to be continuous.)

Quantum mechanics has been proven to accurately describe the workings of this quantum world without exception, from the subatomic particles that make up the constituents of the atom to the most massive components of our galaxies. I list its essential features as follows.[1]

1. Allowed States: Each possible state of an object is described by a complex mathematical function, called a wavefunction, a solution to Schrödinger's equation for that object. The object can usually only be observed in (i.e., occupy) one of these states.
2. Properties: Each wavefunction contains information on the possible physically observable properties of the object (its location, its velocity, its spin, etc.), if it is in the state described by that wavefunction.

3. Overall Wavefunction: Every object or system in the universe is represented by the set of wavefunctions describing its states, and these may be combined into an overall wavefunction.
4. Probability: The probability of the object exhibiting a particular property is calculable for each state from the wavefunction for that state. And the probability of an object being in that particular state is calculable from the overall wavefunction.
5. Measurement: Measuring or observing the state or property of an object determines absolutely the state and property of that object at the instant of measurement. (Upon measurement or observation, a new set of wavefunctions and probabilities is created, and these may change and evolve with the ongoing progression of time.)

Regarding (1), an object can be in any one of the allowed states represented by these wavefunction solutions to the Schrödinger's equation for that object, but in no others. (We recall that Schrödinger's solutions provide the only allowed states for the electron bound to and surrounding the nucleus in the hydrogen atom. It is the transition of the electron from a higher energy state to a lower energy state that releases the quantum of energy [the photon] having precisely the energy [and associated color] given up by the electron in its transition. The distinct colors of light that are observed in the transitions of the electron from one bound state to another correspond precisely to the differences in the discrete energies of the states involved.)

Regarding (2), mathematical operations on the wavefunction can be used to calculate the properties of the object. Properties are more or less likely according to probabilities. (For example, the electron's position is given in terms of probabilities as described in (4) below.)

Regarding (3), everything can be represented by a Schrödinger's equation, and everything is described by the resulting collection of its wavefunction solutions. The scope of inclusion depends on the size of the isolated system being considered. Included may be societies, complex machines, systems of particles, a baseball, a grain of sand, the atom, and so on. Some wavefunctions evolve with time, explaining the way that the objects represented by them can move and spread in space. Some overall wavefunctions represent a collection of stationary spatial states, as we saw for the electron in the hydrogen atom.

Regarding (4), which state an object may occupy and what properties it may have are governed by probability. The probability of movement or events can be determined from the evolving and changing wavefunctions describing the evolution of the states. For example, the probable trajectories of a baseball are given by the evolution of the wavefunctions for the baseball. It is highly, highly, highly probable that large objects such as the ball will follow very nearly the path that would be projected from classical Newtonian physics. But for small submicroscopic objects, the probabilities can be far different from what might be classically expected. (As we have seen, the electron's possible position in the hydrogen atom is described in terms of cloudlike representations of probability with strange symmetries. These have been displayed in Figure 3.8 for five of the electron's lowest energy states. For each state, the probability of the electron being at any particular location is given by the magnitude of the wavefunction at that location, as indicated by the brightness of the cloud. The brighter regions in the clouds indicate a higher probability of the electron being in those regions. The state that an electron may occupy is itself determined by probabilities inherent in the overall wavefunction.)

Regarding (5), in the classical view of physics, the observer is passive. An object is in a particular state and has particular properties. We just observe them, we don't affect them. The quantum world (our actual world) is different. The act of measurement or observation forces but does not determine a particular outcome out of a set of possible outcomes, one state and its properties out of alternative possible contenders. If the same measurement could be made on the same object under the same conditions again and again, the resulting state and properties would each time likely be different and more or less likely to occur according to the probabilities inherent in the wavefunction.

The world works in probabilistic fashion. Quantum mechanics describes this world and operates according to the features listed above. Quantum mechanics is both a framework for understanding and a practical tool for invention. We scientists and engineers accept this quantum view, often approximating very nearly the quantum results for large objects using the simpler-to-work-with classical physics. But what quantum mechanics means, what it says about nature and the role of physics itself, is profound. What follows is a history of the formulation of these ideas and our understanding of them. It all started with Einstein in 1916.

Chapter 6

CLASH OF TITANS—WHAT IS REAL? UNCERTAINTY, ENTANGLEMENT, JOHN BELL, AND MANY WORLDS

THE DEMISE OF CAUSE AND EFFECT, AND OF DETERMINISM?

Part of what Albert Einstein and Niels Bohr discussed when they walked the streets of Berlin in 1920 was the analysis of the Bohr atom that Einstein performed in the summer of 1916. It was in an atmosphere of respect and friendship that Bohr and Einstein began to take opposite sides regarding the validity and implications of quantum theory.

Einstein found that his calculations could predict neither the timing of the transition from one of Bohr's stationary states to another of lower energy, nor the direction in which the resulting light quantum would be emitted. He determined mathematically that transitions might occur at any time, at random, apparently without cause. This meant that, if he trusted his analysis, Einstein would have to give up his classical notion of cause and effect. This troubled him deeply. (Experiments and studies yet to come would add to his concern.)

In the classical, commonly held, view of the world, everything would have a cause and effect. If at any time we could (hypothetically) know everything about the status of every atom in the world, we could in principle calculate everything that had happened everywhere before and predict everything that would happen everywhere afterward. It would be a deterministic world.

Einstein, like most of the rest of the physics community, believed that this was a correct view. However, with his recent analysis he found that even if he knew the present state of the atom and which orbit the electron was in, he wouldn't know if or when it might transition to another orbit. And if he didn't know the direction in which the light quantum would be launched, he couldn't predict what it might cause to happen. Extending this uncertainty to the rest of the atoms in the world, determinism, even in principle, would be gone. His conception (and that of most others) of the way that the universe works would be radically changed.

Einstein felt that this uncertainty and loss of causality just couldn't be, that the emerging quantum mechanics must just be a step on the way to something else, something better that would preserve causality. In contrast, Bohr accepted Einstein's results. He concluded that no predictions ever were or ever would be possible: there never was a predictable "cause and effect," at least on a microscopic scale.

UNCERTAINTY—A FORMAL AND FUNDAMENTAL EXPLANATION FOR INDETERMINACY

In late April 1926, Werner Heisenberg delivered a lecture on his matrix mechanics at the University of Berlin. Einstein subsequently invited the young physicist up to his apartment for an informal discussion on the subject. Heisenberg cites this visit as important to his next accomplishment, particularly a comment by Einstein to the effect that it was theory that indicated what could be observed. Another communication was also a factor. In a letter in October, Wolfgang Pauli described that only momentum but not position could be determined in his analysis (using Max Born's ideas on probability) of the collision of two electrons.

By February 1927, Heisenberg had used his matrix mechanics approach to derive a formal relationship between the uncertainties in various pairs of physical properties. It would become known as the "Heisenberg uncertainty principle." Particular to our discussion, he found that the uncertainty in a particle's position times the uncertainty in its momentum would always be greater than or equal to Planck's constant divided by 4π. (There it is again, Planck's constant.) In symbolic shorthand, we express this as $(\Delta x)(\Delta p) \geq h/4\pi$. Recognizing that the change in momentum, Δp, is approximately m times the change in velocity, Δv, for particles having mass and moving at speeds slow compared to the speed of light, we then have $(\Delta x)(m)(\Delta v) \geq h/4\pi$, or, dividing both sides of the equation equally by m so that they stay in balance, $(\Delta x)(\Delta v) \geq h/4\pi m$. The larger the mass, the larger the number divided into

Planck's constant and the smaller the product of the two uncertainties. (This is often used to explain why we never as a practical matter actually see any uncertainty in the velocity or position of massive objects like grains of sand or baseballs, or rockets or planets. Even a tiny, barely visible, grain of sand is a million billion billion times more massive than the electron, so the uncertainty in the combination of its velocity and its position is very, very, very small. Conversely, the small mass of the electron explains the relatively large uncertainties in position inherent in the spread-out clouds of probability for the location of the electron.)

There is another way to look at uncertainty in position. Grains of sand and larger objects are composed of atoms, and the uncertainty in the size of the atoms is given by the extent of the wavefunctions (represented by probability clouds) of the atom's lower-energy occupied electron states. So the extent of these clouds is a measure of the "fuzziness" and uncertainty in the location of the edges of any larger object composed of atoms. This fuzziness is very, very, very, small in comparison with the overall dimensions of objects composed of billions of atoms. So, we don't as a practical matter deal with uncertainties in the position of large objects, like grains of sand, baseballs, and so on.

Heisenberg thought that the uncertainty principle, forged from matrix mechanics, would support this (his) theory as the preferred formulation of quantum mechanics. He would soon be disappointed.

The ideas in Heisenberg's emerging paper on uncertainty sparked initial criticism from Bohr of some of the physical arguments, and there were ensuing protracted discussions regarding interpretation. Heisenberg argued that the act of observation perturbed the state of an object, so that one really couldn't know that state. It was Bohr's view that the uncertainty would instead and always result from the operation of any observing or measuring apparatus. (Actually, the uncertainty is inherent in the physics and exists whether or not measurement is involved.)

Bohr insisted that waves and particles are just complementary aspects of the same entity, and, to Heisenberg's chagrin, Bohr preferred to explain uncertainty using Erwin Schrödinger's wave mechanics. The argument strained relations between the two men. Heisenberg's paper describing uncertainty as a product of matrix mechanics was published at the end of May 1927. In a postscript he cited Bohr's point of view, that uncertainty was a part of wave-particle duality.

THE COPENHAGEN INTERPRETATION

In subsequent months, Bohr refined his ideas on the dual nature of particles into a principle, which he called *complementarity*: stating that the observation of either wave or particle characteristics (but not both) would be displayed in any situation depending on the choice of information sought by the observer or selected by the nature of the measuring or observational device.[1] Combining this with aspects of Heisenberg's uncertainty, matrix mechanics, and Born's probability interpretation of Schrödinger's wave mechanics, he devised what would later be called the "Copenhagen interpretation" of quantum mechanics. A central tenet is that the act of observation or measurement selects just one of a number of possible physical outcomes for any event, based only on probabilities.

Fig. 6.1. "Youngsters" Enrico Fermi *(left)*, Werner Heisenberg, and Wolfgang Pauli *(right)* at the Como Conference in 1927. (Photograph by Franco Rasetti, courtesy of AIP Emilio Segre Visual Archives, Segre Collection, Fermi Film Collection.)

Bohr presented this Copenhagen view at a meeting of the International Physics Congress in Como, Italy, in September 1927. Among the attendees were Planck, Pauli, de Broglie, Heisenberg, Sommerfeld, Born, and Enrico Fermi. Notably absent were the only two scientists likely to strongly present arguments in opposition to Bohr's theory: Einstein, who would not set foot in fascist Italy, and Schrödinger, who was in the process of moving after being offered and accepting Planck's chair in Berlin. (Planck would retire as professor emeritus.)

COPENHAGEN CHALLENGED— CLASH OF TITANS: SOLVAY 1927 AND 1930

The Como conference was just a prelude to the Fifth Solvay Conference, which was to be held in Brussels later that year. Consistent with diplomatic developments underway at the time, permission was granted by the king of the Belgians to lift the ban on German participation. This time, Einstein and Schrödinger would be present, and the battle over the validity, meaning, and interpretation of quantum mechanics would begin. It would lead in the course of the next eighty-some years to concepts that have strained our classically based sense of credibility further and further.

The weeklong conference started on Monday, October 24, 1927. Hendrik Lorentz chaired the sessions.

> Lorentz is shown in the center of the first row of the photo of attendees, in Figure 1.1, between Einstein and Marie Curie. He shared the Nobel Prize in Physics in 1902 for his work on the influence of magnetism on radiation, but he is perhaps better known for his Lorentz transformations, which were utilized by Einstein in his theory of special relativity. Included are the phenomena of "time dilation" and "length contraction," also known as "Lorentz contraction," which are seen in clocks and objects moving at high speeds relative to an observer. Einstein wrote of Lorentz: "To me personally he meant more than all the others encountered in my life's journey."[2]

The topic of the conference was "Electrons and Photons," but it had been made clear that the conference would be devoted to the new quantum mechanics and related questions.[3] The first three days were spent summarizing and discussing experimental and theoretical progress in areas related to the topic. Einstein and Bohr had been asked to present, but they had refused. They had encouraged and facilitated the work of others but felt that they had not contributed sufficiently themselves. Bohr, Heisenberg, Born, and Pauli (we'll call them the "Copenhagen group") would promote the Copenhagen view, at its core the essential features of quantum mechanics outlined in Chapter 5. Schrödinger, de Broglie, and Einstein would disagree with Copenhagen on more than interpretation. They were seeking a different physics.

De Broglie spoke the next afternoon, describing his seminal suggestion of particle waves, Schrödinger's extension of the idea, and then putting forward the concept of "pilot waves" as an alternative to Born's "probability interpretation" of wave mechanics. Unlike the Copenhagen group, which would have the electron behave as either a particle or a wave, de Broglie envisaged the electron to be a particle "surfing" on a real physical wave that would lead or "pilot" the electron to follow one course or another. (So, in the two-slit experiment, for example, the pilot wave might diffract through the two slits while directing the particle through only one of them.) The concept was attacked from the left and the right: on the one hand by Bohr and company, who wanted to assert their Copenhagen view; on the other, by Schrödinger, who pressed his wave mechanics and the wave nature of the electron. He was still trying to resolve quantum physics with classical ideas. Einstein, who had encouraged de Broglie, remained silent. (De Broglie's pilot wave theory was extended by David Bohm in 1952, but without much acceptance at that time. More on that later.)

On Wednesday morning, Born and Heisenberg shared a presentation of matrix mechanics, Dirac-Jordan transformation theory, Born's probability interpretation of wave mechanics, and the uncertainty principle. In their view, Planck's constant was a result of the basic wave-particle duality of matter. They concluded by stating that quantum mechanics was a "closed theory," complete in itself and no longer susceptible to modification. Though Einstein had been impressed by all that quantum mechanics had accomplished, he was of the opinion that it was just a step on the way to something else, not a closed theory at all. Still, he said nothing.

Schrödinger spoke in the afternoon. He pointed out that there were two theories under the name of quantum mechanics: (1) his wave mechanics, which portrayed objects in a familiar three-dimensional space, and (2) the matrix mechanics of Heisenberg and Dirac, which involved a highly abstract multidimensional space. While the latter would work for hydrogen in three dimensions; helium (with two electrons) could only be represented in a space of six dimensions; lithium (with three electrons) would require nine dimensions, and so on. In his view, the matrix approach could only be a mathematical tool, and ultimately any physical situation described would have to be presented in a realistic three-dimensional space. He also asked how the matrix theory would provide a mechanism for the sudden transition of the electron from one state to another, the quantum jump.

Schrödinger was optimistic that the two theories would ultimately be resolved into one. But he rejected Born's probability interpretation of his wavefunction (mathematics indicating the probability of finding a pointlike particle at any given location). Instead, he proposed that the wave *was* the electron, somehow with a spread-out electric charge. None of the other leading physicists would accept this view, though the community in general found wave mechanics to be a much easier formalism than matrix mechanics for examining the workings of the physical world.

Bohr presented the Copenhagen view. Wave-particle duality is intrinsic in nature within the context of complementarity: the basic tenet is that wave and particle aspects of an object are mutually exclusive in any particular observation. He divided the world into two parts: the micro world, which would be described by quantum mechanics; and the macro world, which would be described in the language of classical physics. The instruments of observation and measurement would lie in the macro world. Reality in the micro world did not exist in the absence of observation. The electron didn't exist at any particular place until some measurement located it. (Heisenberg would go further, to say that it didn't exist at all, anywhere, until the measurement was made.) This would be extended further to say that the electron didn't exist until some cognizant being observed the results of the measurement. There would be no objective preexisting reality.

> Objective reality was essential to Einstein's view of the world. He would point to the moon and contend that it existed whether he was looking at it or not. For him, science was the discovery and understanding of the objectively existing world. And at that time he would not accept uncertainty or that the state or properties of an object would be governed by probabilities. "God does not play dice" (with the universe). He would not accept complementarity, which condemned the wave and particle nature of objects to always be observed separately. For Einstein, there had to be a deeper underlying theory that explained the results of quantum mechanics without sacrificing his fundamental view of a real universe and science's role to discover it.

If Einstein could demonstrate that quantum mechanics was in any way flawed, it wouldn't be considered a closed theory, and the door would be open to seek a better, deeper understanding. His approach to analysis and demonstration in physics often involved hypothetical *gedanken* (thought) experiments. Without actually doing anything in the laboratory, he would test theories by tracing out on paper or in his mind what would happen in various posed experimental tests. He signaled to Lorentz that he would like to speak. The following indented paragraphs describe his use of this approach to attack the claims of the Copenhagen group.

Single-Slit Thought Experiment

> Einstein went to the blackboard and sketched out light particles passing through a single slit in a screen and then impinging on a photographic plate. After emerging from the slit, the photons would spread out [like the water waves emerging from each of the openings in the breakwater in Figure 2.2] according to a wavefunction that would indicate the likelihood that a photon would hit the plate at any point. According to the Copenhagen interpretation of quantum mechanics, at the moment that a single photon hits the screen (at the moment of measurement), the wavefunction for that particle "collapses" to a single spiked peak showing 100 percent probability of the electron being at the point of impact. Einstein asked how the information of the impact could be transmitted instantaneously, faster than the speed of light, to the farther regions of the plate so that the wavefunction in those locations would be triggered to collapse to zero except at the single point of impact. He reasoned that this collapse would either violate "locality" or show the assumption of the collapse of a probability wavefunction (and therefore the Copenhagen interpretation) to be flawed.

Locality had seemed to have been thoroughly tested experimentally classically and was a key element in classical physics, including Einstein's special theory of relativity. It means that objects could only be affected by direct impetus, either by being struck by another object (for example by a photon), or by being moved by the force of an electric, magnetic, or gravitational field. None of these influences, *nothing*, could be transmitted at a speed faster than the speed of light.

Another Explanation

Einstein went on to explain that the observed pattern of interference could result from a classical statistical distribution of the passage of many, many photons. Not only was quantum mechanics wrong, he would say, but the interference results could actually be explained with classical theory. Quantum mechanics wasn't even necessary. A double whammy!!

Bohr had no immediate reply, but later that evening he would respond for the Copenhagen group: The wavefunction was an abstract mathematical entity, had no physical reality, and therefore would not be bound by locality. And he was able to show that Einstein's statistical argument was wrong.

Shifting to a double-slit thought experiment, Einstein next made the case that the position and momentum of a particle could be measured to greater precision than the uncertainty principle would allow.

Bohr sketched out the apparatus and analyzed the experiment in detail, again finding that Einstein's arguments were flawed, and further refining the test to demonstrate complementarity.

In the end, Bohr and his colleagues were able to refute all of the challenges offered by Schrödinger and Einstein. There were still questions, but no flaws in the theory had been revealed. Einstein was still unconvinced. But Paul Ehrenfest, who was unbiased, somewhat of a facilitator, and a good friend of both Einstein and Bohr, would comment that Bohr had pretty much prevailed.

In the next several years, Einstein suffered some health problems, took time off to recover, and began his work to unify the theories of gravity and electromagnetism to produce a "unified field theory." He hoped it would also provide answers to what he still believed to be serious problems with quantum mechanics, and complete the theory. Uncertainty was still on his mind.

Kumar relates, quoting Heisenberg, that most of the action between Einstein and the Copenhagen group actually took place outside of the meeting rooms.[4] In the morning, over breakfast in the hotel where all of the attendees stayed, Einstein would present the group with a new challenge to the uncertainty principle. Details of the challenge would be clarified during the walk to the Institute for Physiology, where the meetings were held. The group would begin to analyze the challenge over lunch. In the early evening, the Copenhagen group would meet to formulate a response, and Bohr would deliver their findings to Einstein over dinner. The entire interaction between Einstein and the group would take place in a spirit of good humor. (Figure 6.2 shows Einstein and Bohr walking together in Brussels seven years later, as they might have walked in 1927.)

Fig. 6.2. Einstein *(left)* and Bohr in Brussels in 1934. (Photograph by Paul Ehrenfest, courtesy of AIP Emilio Segre Visual Archives, Ehrenfest Collection.)

When it came time for the Sixth Solvay Conference in 1930, Einstein delivered what at first appeared to be a fatal blow to the Copenhagen interpretation. The topic of the conference was the magnetic properties of matter. The format and venue of the meeting were the same as it had been three years earlier. There were thirty-four attendees, twelve of them already or later to be Nobel laureates. Again, the dialogue between Einstein and the Copenhagen group would occur between the formal sessions. This time Einstein attacked another aspect of the uncertainty principle. (Remember, if any one aspect of quantum mechanics, including uncertainty, was found to be invalid, then the Copenhagen group would have to admit that quantum mechanics was not a closed theory. It would then be open to the modification and improvement that Einstein felt would be necessary.)

Heisenberg's uncertainty applies to various "conjugate pairs" of physical properties and variables. One such pair, described earlier, is the position and momentum of an object. As we discussed, Heisenberg found that the uncertainty in an object's position times the uncertainty in its momentum would be greater than or equal to Planck's constant, h, divided by 4π, written in physics shorthand as $(\Delta x)(\Delta p) \geq h/4\pi$. Another conjugate pair is energy and time. In this case, the uncertainty in an object's energy times the uncertainty in the time interval for measuring the energy is related to Planck's constant, h, by the formula $(\Delta E)(\Delta t) \geq h/4\pi$.

Einstein confronted Bohr with the following thought experiment.

Suppose, Einstein said, there are many quanta of light, many photons bouncing around inside of a box. (The box is shown in Fig. 6.3, suspended by a spring from a cantilevered beam.)

A shutter over a window in the box is opened and then quickly closed, allowing just one photon to escape out sideways. The shutter can be operated according to a clock, which thus notes the time of the photon's departure with a high level of certainty. The clock is synchronized to another clock located outside, away from the box, so that the observation of time (done on the outside clock) in no way influences the box and its contents. (To this point Bohr is unperturbed. He knows that experimental uncertainty in the measurement of the wavelength, w, of the photon will [by Planck's formula $E = hc/w$, described in Chapter 1] translate into an uncertainty in the energy.) But then Einstein stated that the box would be weighed before the photon is emitted and then weighed again afterward. (The weighing could be accomplished by noting the height of the box against a pointer attached to the stand supporting the beam. The heavier the box, the lower it would drop against the tension of the spring.)

Fig. 6.3. Photo of Niels Bohr's blackboard just after he died in 1962. Apparently he was still thinking of Einstein's photon-from-a-box thought experiment, sketched at the bottom center. (Image from AIP Emilio Segre Visual Archives.)

Bohr immediately realized that uncertainty was in trouble. Here's why:

> Einstein's formula $E = Mc^2$ (from his theory of special relativity) would predict that the box would change its mass because of the energy carried away from it by the photon. By weighing the box before and after the photon's escape, Einstein would be able to measure the change in mass of the box and then (using this formula) calculate the energy of the photon. In this way he would get a precise, exact, measure of the photon's energy without the uncertainty that would be imposed by trying to make direct measurements on the photon itself. There would be zero uncertainty in the calculation of energy, and therefore zero uncertainty in the product of this uncertainty times whatever uncertainty existed in the measurement of time. (Zero times anything is zero.) The uncertainty principle would be violated since it predicts that the product of these uncertainties must always be greater than $h/4\pi$. Quantum mechanics as the Copenhagen group envisioned it, including uncertainty, would be shown to have this flaw and would therefore be considered to be incomplete.

Bohr conferred with his colleagues, but none could find anything wrong with Einstein's argument. They were relatively unconcerned, confident that a way out would be found. Bohr was still greatly concerned. In Bohr-like fashion he sought to examine the details of measurement, sketching out a simple concept of the box with its clock and shutter, the box hanging by a spring like a grocers scale with a pointer to measure its displacement as a way of indicating weight.

Bohr struggled. Eventually, in the early hours of the morning, he found what he was looking for. Einstein had invoked relativity. He would out-Einstein Einstein. It was very subtle.

> Bohr noted that light, photons, would need to shine on the box and the pointer in order for its location to be read. The photons, each having momentum, would jostle the box in a small way. Its position, and therefore the mass and energy of the box would be uncertain.
>
> And the measurement of time would also be uncertain: The more accurate the reading made of the pointer's position, that is, the more accurate the reading of the mass change of the box of light and the associated energy of the single emitted photon, the greater would be the required number of these externally generated photons and the greater the random jostling that they would produce. Einstein's theory of general relativity, the same theory that successfully predicted the bending of the path of light as it passes the sun, also requires that time slows and clocks tick more slowly if they are moving in a gravitational field. So, if a change in mass is measured by weighing it under the force of gravity as Einstein suggested, then Einstein's clock in the box would (because of random movement in the gravitational

field) run in a random way more slowly than the initially synchronized clock viewed by the observer.

Time would be uncertain, and the more accurate the measurement of mass and energy, the more uncertain time would become. This was exactly what Heisenberg's uncertainty principle said of the conjugate quantities of time and energy!

Einstein had no rebuttal. He still maintained that quantum theory was incomplete, but from this time on he would try to demonstrate this head-on rather than indirectly by attacking uncertainty.

Most of the leading physicists taking positions in the various universities throughout Europe were directly or indirectly a product of Bohr's Institute. His constant invitation to visit, share, and collaborate had produced a cadre of believers whose students would now carry on with the Copenhagen interpretation. Einstein in his Institute in Berlin had preferred to work alone. He stood out (with Schrödinger) as a reactionary, nearly lone holdout still seeking to demonstrate an objective reality.

EINSTEIN GAINS THE UPPER HAND? ENTANGLEMENT AND THE (EPR) PARADOX

In 1935, Einstein launched another attack. By this time he was working at Princeton in the United States. Together with colleagues Boris Podolsky and Nathan Rosen, he devised yet another thought experiment and delivered the findings in a paper published in May in the American journal *Physical Review*, titled "Can Quantum Mechanical Description of Physical Reality Be Considered Complete?" Their answer was "No!" The experiment involved an aspect of quantum mechanics later to be labeled by Schrödinger as "entanglement." The EPR (Einstein, Podolsky, Rosen) paper was concerned with entangled properties of momentum and position, but we illustrate entanglement here because of their relevance to experiments to be described later on, by considering photons with the entangled property of polarization.

As we will discuss later on, entanglement allows the possibility of superpowerful quantum computers, quantum encryption, and even one type of teleportation.

Objects are entangled if they are both part of the same wavefunction. A good example of entangled particles is the pair of photons that can be induced to be emitted within nanoseconds of each other from a calcium atom.

All photons are electromagnetic in nature and, thinking classically of waves to understand definitions, are polarized as described in Appendix A. Their tiny electric fields alternate in time and space like a rope whipped up and down, left and right, or at some odd angle in between. Whatever the angle, the oscillations of electric field lie in only one plane, vertical, for example, as depicted in Figure A.1(c) of the Appendix.

The two photons from the calcium atom are constrained by the physics of the emission process of our example to have identical polarizations and to travel in opposite directions. Their polarizations can be at any angle. If a measurement of polarization is made on either photon, let's say by passing it through a polarized filter, it will either make it through or not. If the filter were set to pass vertically polarized photons, and our photon were polarized in the vertical plane, it would pass through. If it were polarized in the horizontal plane, it wouldn't make it through. If it were polarized at some odd angle, it will either entirely make it through the vertical filter or not at all, with some probability of each result depending on the angle of its initial polarization relative to the vertical. The photon doesn't partly make it through. It either assumes the polarization that passes the filter or the orientation 90 degrees from that and doesn't pass. This goes back to quantum mechanics and the particle and quantum nature of photons: they don't get split.

The key thing about entangled photons is this: whatever the polarization measured on the first photon, that same polarization will be measured on the second photon, even though the result on the first photon is based entirely on probabilities as quantum mechanics would expect, and even though the two photons may be far, far away from each other. Both the photon and its entangled partner choose through this one polarization measurement (on the first of the two photons) the same one of two possible results: "pass" or "don't pass" through similarly oriented filters, regardless of the polarization angle of these filters. Once the pass/don't-pass measurement is made on the first photon (once it is shown to have either the polarization of the first filter or the opposite polarization), there is no probability involved in regard to the polarization of the second photon: it will be the same as the polarization measured on the first photon.

EPR examined in thought what would happen with measurements on entangled particles that had been allowed to travel far, far away from each other. We illustrate the EPR argument with our two photons.

Quantum mechanics as envisioned by Bohr would say that the photons' entangled polarization would not be determined until a measurement was made on one of the photons by trying to pass it through a polarized filter: either it would pass or not and be in a "pass" state, or not, as a result of the measurement. A similar measurement made immediately afterward on the other, far-away, photon would always show it to be in exactly the same polarization state as the first photon: that is, it would pass through a similarly oriented polarized filter, or not, in exact correspondence with what happened with the first photon. Its polarization would also be determined by the measurement on the first photon.

EPR would ask how this is possible. How did the second photon know the randomly determined polarization of the first photon, so that it could assume its same polarization state?

Either (1) the experiment would violate locality (the requirement that events are caused by contact or the transmission of forces no faster than the speed of light) by transferring information to the second photon instantaneously, what Einstein would call "spooky action at a distance." (The transfer of forces or information faster than the speed of light was not allowed according to any prior classical experiments or the much-trusted theory of special relativity, a part of classical theory). Or (2) the photons had the same polarizations all along, in which case quantum mechanics was lacking and incomplete as a theory because it should have been able to describe the definite preexisting polarization of the pair of photons—no probabilities involved.

Bohr had thought that he and the others of the Copenhagen group had disproved Einstein's classical sense of objective preexisting reality through their Copenhagen view of quantum mechanics that said that a subjective measurement would force an outcome based on probabilities, that the properties measured in particles would not only be found but actually determined at the instant of measurement. EPR turned the argument around. They asked, why assume that quantum mechanics is right? Since nothing could tell the second particle (faster than the speed of light) what state it should be in, and its state is clearly linked to that of the first particle, then both particles had to be in the measured state all along. Something had to be wrong or incomplete about quantum mechanics since it was not able to predict this prior state. Perhaps there were as yet unknown "local hidden variables" involved in the theory that would describe this preexisting state of the particles. The Copenhagen interpretation of quantum mechanics or quantum mechanics as a complete theory was threatened.

This was not an idle debate or an argument of philosophy, though it seems philosophical. Here were the best of scientific minds trying to determine the nature of our physical world based on experiments that had been or could be performed.

Bohr dropped everything else that he was doing and spent the next six weeks working night and day to analyze and draft a reply to the EPR paper. When he was done, he wrote a paper that bore the same title as that used by EPR: "Can Quantum Mechanical Description of Physical Reality Be Considered Complete?" But *his* answer was "Yes."

The paper was not well written. Bohr had been unable to find any error in the EPR thought experiment and could only argue that EPR's was not a strong-enough case. Paul Dirac initially thought that Einstein had disproved quantum mechanics. In the years that followed, the dispute became essentially a standoff. Which view one took became a matter more of belief than science. But quantum mechanics worked to describe the submicroscopic world, and because of that most physicists, including Dirac, still tended to favor Bohr.

Much later, in 1949, Bohr would suggest that entanglement meant that the particles could not be considered as individual objects. He would reason that they were essentially one object, linked by a single wavefunction: a measurement on one particle essentially constituted a measurement on both particles, simultaneously bringing about the entangled properties in both. No need for faster-than-light transmission of cause and effect.

Fifteen years later still, a theorem was derived that suggested an experiment that would resolve concretely the dispute as to whether quantum mechanics would prevail as a complete theory. The experiment would be a resolution between a world based on probabilities and a world based on objective reality. The theorem came to be called "Bell's theorem" or "Bell's inequality." We'll learn more of that later.

SCHRÖDINGER'S CAT AND "MANY WORLDS"

The EPR paper prompted a series of letters between Einstein and Schrödinger. In one of these, Schrödinger (perhaps anticipating Bohr's later observation) would comment that the act of measurement should immediately break entanglement and leave the properties of the second particle to be independent of those of the first. In another, addressing

Copenhagen's ad hoc separation of micro and macro phenomena, Einstein sought to demonstrate the absurdity of the Copenhagen interpretation as it might apply to larger objects. This triggered a subsequent paper, published by Schrödinger, in which he went a step further, describing in a paragraph what is referred to as the "Schrödinger's cat paradox" (which is more known to the general public than is Schrödinger himself or his key theoretical work on wave mechanics).

Schrödinger agreed with the Copenhagen interpretation that all of the information on the probabilities of future events is included in his time-dependent set of mathematical wavefunctions for the state of an object or system. But he disagreed with the part of their interpretation that required that the state of an object can only be determined by observation or test by a cognizant observer. In his own words (but translated from the original German and presented by Kumar), Schrödinger wrote as follows:

> A cat is penned up in a steel chamber, along with the following diabolical device (which must be secured against direct interference by the cat): in a Geiger counter there is a tiny bit of radioactive substance, so small, that perhaps in the course of one hour one of the atoms decays, but also, with equal probability, perhaps none; if it happens the counter tube discharges and through a relay releases a hammer which shatters a small flask of hydrocyanic acid. If one has left this entire system to itself for an hour, one would say that the cat still lives if meanwhile no atom has decayed. The first atomic decay would have poisoned it. The wave function of the entire system would express this by having in it the living and the dead cat (pardon the expression) mixed or smeared out in equal parts.[5]

Kumar continues:

> According to Schrödinger and common sense, the cat is either dead or alive, depending on whether or not there has been a radioactive decay. But according to Bohr and his followers, the realm of the subatomic is an *Alice in Wonderland* sort of place: because only an act of observation can decide if there has been a decay or not, it is only this observation that determines whether the cat is dead or alive. Until then the cat is

consigned to quantum purgatory, a superposition of states in which it is neither dead nor alive.[6]

Twenty years later, a solution would be proposed to the problem exemplified by the Schrodinger's cat paradox. As you will come to see, it would involve the concept of the simultaneous existence of "many worlds."

In the interim, our one of these many worlds would sink once again into war.

NAZI GERMANY, NUCLEAR PHYSICS, AND THE BOMB

Between 1928 and 1930, Hitler's National Socialist party moved from just 12 seats in the Reichstag to 107 seats, making it the second largest party in Germany. What precipitated the change was the crash of the financial markets on Wall Street.

American banks, in trouble, demanded repayment of short-term loans which had stimulated the German economy. German unemployment, which was at 1.3 million workers in 1929, rose to three million in 1930. A year later, Germany was in deep depression and political upheaval. Hitler exploited a simmering anti-Semitism by blaming Germany's problems on the Jews. He was appointed chancellor in January 1933. State-sponsored Nazi violence began when the Reichstag was set on fire just one month later. More than a quarter of Germany's roughly sixty-five million people would vote for the Nazi Party in the Reichstag election in March.

Five days after the Reichstag fire, Einstein, who was lecturing at Caltech in the United States (and had arranged to spend a few months per year at the newly formed Institute for Advanced Studies [IAS] at Princeton), decided not to return to Germany. He stated publicly that he wouldn't live in a country where basic freedoms were restricted. He was vilified in the German press. In May "un-German" and "Jewish-Bolshevik" books and documents were looted from libraries and bookstores and burned in every university town in Germany. Included were the works of Brecht, Freud, Kafka, Marx, Proust, Zola, and Einstein.

The Law for the Restoration of the Career Civil Service had passed in April. Civil servants not of Aryan origin were to retire. Universities were state institutions, and by 1936 over 1,600 scholars had left their posts. A third were scientists. (Twenty of these either had been or would be awarded Nobel prizes.) Included were a quarter of all members of the physics community and half of the theorists.

Schrödinger didn't have to leave Berlin, but he did so out of protest. Protest by the rest of the German physics community was feeble at best, but scientists in other countries set their organizations to helping out. Ernest Rutherford chaired the Academic Assistance Council in England, which served as a clearinghouse in finding positions. Bohr's institute became a staging post, and he and his brother helped to set up the Danish Committee for Support of Intellectual Workers in Exile. The better-known physicists would be able to find places outside of Germany. It was much more difficult for the rest.

Born didn't have to leave because of a "grandfather" clause in the law, but he felt uncomfortable among unsupportive colleagues. He left Germany to lecture for three years at Cambridge and then to take the chair of natural philosophy at the University of Edinburgh in Scotland. His institute, the "cradle of quantum mechanics"[7] in Göttingen, essentially ceased to exist.

In the late 1920s, physics was mainly concerned with the quantum mechanics of the electrons surrounding the nuclei of atoms. During the 1930s, experimental work would provide information toward an understanding of the nucleus itself. Bohr had visited Einstein at Princeton on several occasions. When he returned in January 1939, he brought news that European laboratories had discovered fission—that heavy elements might break apart into smaller elements, releasing large amounts of energy and neutrons in the process. These might trigger the fission of additional atoms in a chain reaction that could be used to build an atomic bomb. Bohr also reported that Germany had stopped the sale of uranium ore from the mines that it now controlled in Czechoslovakia. (Approximately 0.7 percent of uranium ore is the fissile isotope uranium-235.)

Under the circumstances, Einstein suspended his pacifist views and in August 1939 wrote a letter to President Roosevelt suggesting that he look into the possibility of building an atomic bomb. In September, Germany attacked Poland. In March 1940, Einstein wrote a second letter to Roosevelt, stating that German interest in uranium had intensified and that much work was being carried out in Germany in secrecy. (He didn't know that Werner Heisenberg was in charge of the German project to develop an atomic bomb.)

In April 1940, Germany occupied Denmark. Bohr stayed in Copenhagen, hoping that his reputation would provide some protection to others who were there. In September 1943, Hitler ordered the deportation of Denmark's eight thousand Jews. Thanks to the Danish people, who hid them in their homes, in hospitals and churches, only three hundred were rounded up. Bohr, whose mother had been Jewish, escaped with his family to Sweden and then was flown to Scotland as a passenger on a British bomber. From there he traveled to the United States. After a brief stop in Princeton, where he met with Einstein and Pauli (who was at the Institute for Advanced Studies at the time), Bohr traveled on to Los Alamos, New Mexico, to take his place among those working on the atomic bomb. He worked under the alias "Nicholas Baker." The United States hadn't started a serious effort toward the bomb until 1941. When it did, it was codenamed "the Manhattan Project."

BELL'S INEQUALITY, EXPERIMENT, AND THE NATURE OF REALITY RESOLVED

After the war, Bohr was made a permanent nonresident member of the Institute for Advanced Studies at Princeton. He and Einstein met there in 1946, again in 1954, and several times in between. On all occasions they again discussed quantum mechanics and its implications. Heisenberg also met with Einstein in 1954, stopping by as part of a lecture tour. Neither man could sway Einstein from the latter's belief that quantum mechanics was incomplete, just a stepping-stone toward a broader theory that would describe a "real" and "local" world.

Einstein died of a ruptured aneurism the next year, apparently still thinking about the development of a unified field theory that might include a resolution of the problems he saw in the implications of quantum mechanics. Bohr died of a heart attack in 1962, having sketched Einstein's light-box thought experiment on his blackboard the night before (see Figure 6.3), apparently still carrying on in his mind the debate with his old friend and adversary. In 1964, a young Irishman would construct a theoretical proof that showed a way to resolve their dispute.

John Stewart Bell was born in 1928 in Belfast, into a poor family "descended from carpenters, blacksmiths, farm workers, laborers and horse dealers."[8] His family managed to send him to a technical high school; he was the only one of the four children afforded a secondary education. Luckily, he found work as a technician at Queens University, was recognized for his drive and talent, was given a small scholarship, and with diligence graduated in 1948 and 1949 with degrees in experimental and mathematical physics, respectively. He went to work in England for the United Kingdom Atomic Energy Research Establishment, married fellow physicist Mary Ross, and then went on to earn a doctorate at the university in Birmingham. In 1960, he and Mary moved to Geneva, where he would work at the European nuclear physics research facility, CERN, there to help in the design of particle accelerators.

Fig. 6.4. John Stewart Bell at CERN in 1982. (Image courtesy of CERN, the European Organization for Nuclear Research.)

Though his job was to design and build scientific apparatus, Bell had from as early as 1945 been following theoretical works regarding the implications of quantum mechanics. We know today that quantum mechanics, with all of its uncertainties and probabilities and nonlocal character, has succeeded without fail in describing our observable universe. But this is not to say that the Copenhagen interpretation at the time would prevail or that quantum mechanics would be viewed as a complete theory. In the following indented paragraphs we examine how Bell began to explore those interpretations that existed at the time.

> In 1932, the brilliant young mathematician John von Neumann wrote a book on the mathematical foundations of quantum mechanics that became the definitive reference on the theory. In this book, von Neumann appeared to prove that quantum mechanics could not be reformulated to include "hidden variables" that would allow the objective reality that Einstein sought. Nevertheless, some twenty years later, David Bohm, who had been a student of Robert Oppenheimer at Berkeley, constructed a credible hidden-variables theory related to de Broglie's earlier

work describing the electron as a particle "surfing" a pilot wave. Because of von Neumann and other factors, and because quantum mechanics based on the Copenhagen interpretation had become so firmly entrenched, Bohm's work had been pretty much ignored. But not by Bell. Eventually, Bell examined von Neumann's work and found his argument about hidden variables to be flawed.

Stimulated by Bohm's work, Bell in 1964, in a sabbatical year away from CERN, decided to resolve the dispute between the accepted Copenhagen view of quantum mechanics and Einstein's argument that it either violated locality or was incomplete. Einstein concluded the latter. Bell tended to sympathize with Einstein's position and set out to demonstrate the possibility of a "local hidden variables" (LHV) theory alternative to quantum mechanics.[9] Based on a theorem that he derived, Bell suggested an experiment to test between LHV and quantum mechanics by measuring the correlations between the passage of particles with entangled spins through differently oriented filters (i.e., pass/pass; or don't pass/don't pass). Each set of particles tries to pass through filters set at different angles. That is, one particle passes through (or not) a filter set at a particular angle, while its entangled partner passes through (or not) a filter set at a different angle. If the correlations fell into a certain range, then Bohr would be right and Einstein would be wrong. It was a difficult test, not accomplished properly until nearly fifteen years later, and then with polarized entangled photons rather than spins. I describe the test as follows.

As with our earlier consideration of entanglement, because we can relate it to the definitive experiments that were eventually performed, we describe Bell's experiment using a pair of polarized photons instead of a pair of particles with oriented spins. (I refer you to Figure A.1(c) and related discussion in Appendix A for a definition of polarization.) The pair of entangled photons is emitted through the two-stage transition of an electron from an excited state in a calcium atom. The two photons travel in opposite directions. Upon measurement, both photons will have the same polarization, but what that polarization might be is unknown. The two photons are each passed through a polarizing filter, and each filter may be oriented at a different angle to the vertical than is its counterpart for the other photon.

Correlated results are achieved if either both photons pass through their filters or both do not pass through their filters. A lack of correlation results if one photon passes the filter and the other doesn't. For LHV, where the two photons are assumed to already have a particular matching polarization when they are emitted, the probability of correlation in the measurements on the two filters is calculated never to be less than one in three (1/3), regardless of how differently the detecting filters may be oriented. (This is "Bell's inequality": for LHV to be possible, the probability of correlation must be equal to or greater than one in three, about 33 percent.) For quantum entangled particles, where the polarization of both particles is not determined until the first measurement on either of the two particles is made, the probability of correlation in the measurements is calculated to be as low as one in four (= 1/4 or 25 percent).

Bell suggested that many measurements be made at each of many different relative orientations of the two filters, generating a curve of probabilities to see what the minimum probability would actually be. If it were found to be in some range not

> allowed by LHV, then deriving any LHV theory, including one obtained as a modification of quantum mechanics, would be impossible. For our case of the polarized photons, this meant that some combination of filter orientations would need to be found such that measurements would show a probability of correlation significantly below 33 percent.

Ever more careful and precise sets of measurements were performed to test Bell's inequality, starting with John Clauser and his group at Berkeley in 1969, using photons as described in the indented paragraphs above. Clauser's work found in favor of Bohr's nonlocal quantum mechanics, but it wasn't definitive. In three sets of experiments, similarly with photons, Alain Aspect (pronounced "AHS-pay") and his group at the Université Paris–Sud in Orsay in 1981 and 1982 found with a high degree of accuracy that the polarizations of the photons could be correlated close to the 25 percent minimum expected from quantum theory and clearly less than the 33 percent minimum required for LHV.[10] The conclusion is, with a very, very, very high degree of confidence: Einstein was wrong! In Orzel's words: "the vast majority of physicists agree that the Bell Theorem experiments done by Aspect and company have conclusively shown that quantum mechanics is nonlocal. Our universe cannot be described by any theory in which particles have definite properties at all times, and in which measurements made in one place are not affected by measurements in other places."[11] Said in another way by Kumar: "Aspect's team and others who tested Bell's inequality ruled out either locality or objective reality, but allowed a non-local reality."[12] Einstein would either have to give up objective reality or accept "spooky action at a distance" (and would probably have chosen the latter).

INTERPRETATIONS OF QUANTUM MECHANICS

The confirmation of Bell's theorem ruled out *local* hidden-variable interpretations and posthumously settled the dispute between Einstein and Bohr. However, as of 2016, thirteen distinct interpretations have been put forward of how quantum mechanics works).[13] I note the three mainstream types of interpretations here.[14] One type is the de Broglie/Bohm hidden-variables interpretation (mentioned earlier), which involves additional math to describe a particle surfing a pilot wave. It includes hidden variables that are not local, and so is not ruled out by the Bell experiment. In this interpretation there is a deterministic flow of events, but the determinism in the flow is not observable.

Another type is labeled as "collapse theories," which, including and like the Copenhagen interpretation, requires that an object be in a superposition of possible states until observation or measurement causes the collapse of the overall wavefunction into just one state and set of properties. As noted earlier, the absurdity of this for large objects was illustrated by Schrödinger's cat paradox, where the cat was deemed to be both dead and alive until observed.

In 1957, a simple solution to the problem of collapse was proposed. Hugh Everett III, a graduate student at Princeton, suggested that our world splits at every event that allows alternative outcomes. All possible outcomes are realized, each in a separate ongoing world. Each of these worlds is split again at the next event, so that there is an ever-branching tree of separate realities. This latter became known as "many worlds," variations of which make up our third "mainstream" type of interpretation (though, to many, it still seems far-fetched).

Everett was able to show that his approach would yield all of the practical quantum calculations and results provided by the Copenhagen interpretation. As practical physics, it was the same. But at the time that his idea was proposed, the Copenhagen interpretation was well entrenched, and Everett's suggestion went essentially unnoticed for almost ten years. Then cosmologists begin to have trouble with the Copenhagen idea that an objective observer would be required to cause the collapse of the wavefunction into one of its possible outcomes. Where was the outside observer to cause the collapse of the set of wavefunctions describing the universe, so that we have the universe that we occupy? Better to believe in the existence of multiple branching universes requiring no observation and no collapse.

Polls

Kumar cites a poll taken at a physics conference in Cambridge in 1999.[15] Ninety physicists were asked about interpretation. Four favored the Copenhagen view. Thirty preferred the "many worlds" concept. Fifty stated "none of the above" or "undecided." Another poll carried out by

Max Tegmark at the Fundamental Problems in Quantum Theory conference in August 1997 cited a 17 percent vote for many worlds, about the same as the 18 percent polled in a Quantum Physics and the Nature of Reality conference in July 2011, which also reported 42 percent for the Copenhagen interpretation.[16] So, even to this day, the jury is out regarding which of the various interpretations are likely to be correct and which of the corresponding views of our physical world may be correct. (Hopefully, someday the correct interpretation will be determined by the experimental check of some distinguishing theory, rather than through the expression of a preference by a vote.) Fortunately, quantum mechanics works regardless.

DECOHERENCE—WHY OUR MACROSCOPIC WORLD SEEMS TO BE CLASSICAL

The physical process of "decoherence" is particularly relevant to the many worlds interpretation, but it relates to all of the other interpretations as well. It answers some questions about what happens upon observation or measurement, and it has particular relevance to quantum computing, encryption, and teleportation (as described in Chapter 8). To understand decoherence, we first need to understand what is meant by "cohere."

Recall from Chapter 5 and the essential features of quantum mechanics that all states of an isolated physical system are described by the wavefunction solutions to Schrödinger's equation for the system, and that these wavefunctions may be incorporated into an overall wavefunction. Well, these wavefunctions are linked together and describe the linkage of all of the constituents of the system. The maintenance of these linkages as the wavefunctions of the system evolve is called *coherence*.

Sustained coherence occurs when particles stay entangled as they move apart over long distances. It is responsible for the interference of a particle with itself as it is shot to pass through two slits in a barrier, as described for electrons in relation to Figure 3.4. In short, it is what distinguishes quantum phenomena from the classical; coherence is a central feature of our quantum world.

Decoherence is the modification of the wavefunction so that the linkage is weakened between otherwise-separate entities in the system and possible resulting states of observation. The phase relationships within and between wavefunctions are shifted. Decoherence occurs through the interaction of the otherwise-isolated system with the broader environment, for example when molecules in the air destroy the linkage between entangled photons as they travel miles through the atmosphere (in a demonstration of quantum encryption, as described in Chapter 8). And it occurs when the billions of atoms in measuring devices perturb the wavefunction of the particles or system that they are observing or measuring. Orzel points out that "it's a real physical process compatible with any of the interpretations (of quantum mechanics)—but it's particularly important to the modern view of many-worlds (which is sometimes called 'decoherent histories' as a result)."[17]

The many worlds interpretation requires that I in my universe cannot see and am unaware of the me observing the different outcome of an event in another universe. But because both of these universes have evolved from the same set of wavefunctions, they should be linked and interfere with each other, in much the same way that the electron interferes with itself as its wavefunction somehow senses the other slit in the double-slit experiment described in and around Figure 3.4. The argument is that this linkage doesn't happen, the many particles within each of these universes are continually perturbing their wavefunctions, shifting their phase relationships, so that they no longer cohere or interfere. The two universes then behave as if they are separate, isolated systems.

Decoherence even places the Copenhagen interpretation in a more favorable light, explaining the difference between the interactions of microscopic and macroscopic systems, why macroscopic objects don't display quantum behavior. The perturbations of billions of atoms in macroscopic objects causes decoherence, so that only the most likely of possible separate results is visible, in the same way that attempting to observe which slit the electron went through destroys interference and yields only the results expected if the electron had gone through one slit. To explore this subject further, I recommend that you read from Chad Orzel[18] or Brian Greene.[19]

Soon, in Chapter 8, because we have already laid some of the groundwork for it in our discussion of the hydrogen atom and polarized photons, we'll take a peek at a few present and future related applications of quantum mechanics. (Much more of future applications will follow in Part Five of this book, beginning with Chapter 18.) But now, let's first examine the meaning of science in the context of all of the new ideas that we have been describing.

Chapter 7

WHAT DOES IT ALL MEAN?—QUANTUM MECHANICS, MATHEMATICS, AND THE NATURE OF SCIENCE

ON PHILOSOPHY, NATURE, AND THE ROLE OF MATHEMATICS

Since ancient times, our level of understanding has grown along with the level of the mathematics that has been developed and employed. Five hundred years ago, Galileo wrote:

> Philosophy [i.e., physics or natural philosophy] is written in this grand book—I mean the universe—which stands continually open to our gaze, but it cannot be understood unless one first learns to comprehend the language and interpret the characters in which it is written. It is written in the language of mathematics, and its characters are triangles, circles, and other geometric figures without which it is humanly impossible to understand a single word of it; without these, one is wandering around in a dark labyrinth.[1]

Newton, by 1687, had developed a differential calculus that he used to describe the motion of the planets.[2] And even in a much more modern time it was Schrödinger who indicated the need for yet more math. In his notes of December 27, 1925, on the eve of his development of wave mechanics, he wrote:

> At the moment I am struggling with a new atomic theory. If only I knew more mathematics! I am optimistic about this thing and expect that if I can only . . . solve it, it will be *very* beautiful.[3]

Through the mathematics of quantum mechanics we have a view into the workings of the quantum world on a submicroscopic scale. What we see is strange to us, and beautiful with its mathematical patterns of uncertainty and probability yielding accurate descriptions of atoms, the building blocks of the world of our senses. And, as you will see, quantum mechanics even explains aspects of black holes, the centers of our galaxies. But is this quantum world only in our minds (because the view provided by quantum mechanics *is* mathematical and therefore a construct of the human mind)?

Consider: Mathematics is just a way of using logic with the aid of symbols. If protons and electrons *do* have electrical charges and attract each other as we have observed, and if the electron *can* be accurately modeled mathematically as diffuse in form and attracted to the nucleus as Schrödinger has done, then the solutions to his mathematical equations are just logical conclusions.

The quantum results of Schrödinger, Heisenberg, and Dirac so well describe this world that we tend to believe that theirs are functionally correct models of the electron and the atom. But the quantum world, including the galaxies, would still exist even if we were not here to make these mathematical models. So it seems that nature itself must be inherently logical, and if there are charged electrons and protons, then the quantum atomic behavior that we observe occurs in a very natural, logical way. Our mathematics just describes it.

HYPOTHESIS, THEORY, LAW, AND CORRESPONDENCE

This book describes parts of quantum theory and the results of its more formal mathematical structuring as quantum mechanics, most specifically as applied to the atom. *Theory* is defined as a well-tested and proven set of ideas that accurately explains the physical world. Theory by this definition has otherwise sometimes been called a *law*. Theory is well proven, as compared to untested postulates and ideas, which are scientifically labeled as mere *hypotheses*.

A successful theory must be able to predict the occurrence or observation of things or happenings not previously seen. As noted earlier in this book, quantum theory in its broader context, including quantum mechanics, is the most proven, predictive, and successful theory in the history of science.[4] Quantum theory has become accepted as a broad theory that is valid over the entire range from the macroscopic (large objects) through the microscopic, the submicroscopic, subatomic, and even subnuclear. And it has far-reaching implications, rendering invalid the previous classical (deterministic)[5] view of what we can observe of our world.

It is not that scientists have capriciously revised their thinking since classical times. Nor that quantum theory was suddenly adopted as a popular new set of ideas. Rather, it was on the basis of more and more experimental evidence that didn't fit with the classical theory that finally scientists had to give up on the classical and seek out something new in explanation. And even when many leading scientists accepted this new quantum theory, there were many who found it just too radical a departure from classical thinking. So the development of quantum theory (including its mathematical formalization as quantum mechanics in 1925) was truly a *scientific revolution* in the sense described by Kuhn.[6]

As outlined in the preface and early chapters of this book, the classical physics of Newton, Maxwell, and others (which preceded quantum theory) that seemed to so well describe the physical world, began not to fit with newer discoveries starting around 1900. In particular, much of classical physics simply did not explain the submicroscopic world. And we now know that classical physics is also *conceptually* wrong even as it applies to large objects—say, objects the size of a grain of sand or the planets.

However, despite its failures in the realm of the submicroscopic, classical physics does provide a very good *approximation* to the results of quantum mechanics for large objects like grains of sand or the planets.[7] We call this agreement for large objects *correspondence*, and, because classical physics is easier to use than quantum mechanics, we continue to apply classical physics theory in those many parts of science and engineering where the approximation is sufficiently accurate. In a similar way, should newer, more broadly applicable theories be developed, quantum mechanics and relativity will not simply be cast aside. They will still apply, in another, newer region of correspondence.

Next, before we get a look (in Part Three) at our quantum and relativistic world from the tiniest of fundamental particles to the galaxies, we visit another exciting example quantum application, one that is just emerging with great potential.

Chapter 8

APPLICATIONS— QUANTUM COMPUTING, CODE CRACKING, TELEPORTATION, AND ENCRYPTION

In Chapter 6 we discussed entanglement in the context of Bell's inequality. We examine now a few applications that make use of this strange phenomenon.

INTRODUCTION TO CHAPTER 8

The development of quantum computers is driven in part by an ongoing commercial, governmental, and military need to transmit information in secret, and a corresponding need in the State Department and the military to be able to break into and read coded information. It is not just that the quantum computer can operate faster than a classical computer *for some applications, but rather that* it can operate in an entirely different manner by simultaneously addressing the problem at hand through computations in multiple paths, all using the same circuit elements. The same quantum properties promise faster computation for many other applications, including accurate simulations of physical situations in our (quantum) world.

While quantum computers will allow the cracking of classical codes, quantum properties provide new methods of encryption, codes that would seem as though they can in no way be broken. And any messages encoded in this manner cannot be tampered with without the sender and receiver knowing about it. These same properties make quantum copying or cloning impossible, while in principle allowing a form of teleportation, something not even theoretically possible by any classical means.

Please realize that I will be selective in my description of the topics in this chapter. I provide only enough for basic understanding and a sense of the exciting opportunities that lie ahead. For an entertaining and more comprehensive consideration of the history of development in quantum computers, encryption, and related physics, I suggest that you read *Computing with Quantum Cats*, by John Gribbin (Reference Z). For a somewhat-deeper examination of the essential elements of classical and quantum information theory, computers, cryptography, cloning, and teleportation, I similarly suggest lectures 19 through 22 of *Quantum Mechanics: The Physics of the Microscopic World*, by Benjamin Schumacher of Kenyon College (Reference Y). And, for a good recent overview not only updating broadly the development of quantum-computer elements but also indicating the hardware and software steps necessary to produce functioning quantum computers, I would recommend the YouTube video presentation by Krysta Svore of Microsoft, shared on October 23, 2014.[1] To provide a context and define basic terminology, I start now with an examination of what can be done with classical information.

CLASSICAL INFORMATION IN BINARY DIGITAL FORM

I find it convenient to borrow a description from Schumacher's lecture 22. He starts out by asking us to remember the quiz game Twenty Questions, noting that every yes or no answer to the questions asked represents a single "bit" of information—in all, twenty bits. (The answers are *binary*: that is, they have only two possibilities: yes or no. These answers may be represented by the digits 1 and 0, hence the term "bit," for "binary digit.") Schumacher points out that the various combinations of these twenty bits, if applied toward guessing a particular word,[2] are sufficient to sort through all but a few of a million possible words in the English dictionary. Said another way, all of these million words can be represented by different combinations of these twenty bits.

This seems rather incredible from twenty pieces of information, but we can see how this may happen considering just a few bits to start with. The first bit allows either of two possible answers to the first question, yes or no, represented by 1 or 0. For each of these, the second bit allows 1 or 0, so that there are four combinations in all. For each of these four combinations there are two possible answers to the third question, making eight combinations in all. Then there will be sixteen combinations with the fourth bit and 32, 64, 128, and so on, as more bits are added. The number of combinations goes up pretty fast. We write the number of combinations in mathematical shorthand as 2^n, where the exponent *n* represents the *n*th bit, and the expression means that 2 is multiplied by itself *n* times. And so, we say that the number of possible combinations increases exponentially. So, in the game of Twenty Questions, we find that the number of bits, 2^{20}, is exactly 1,048,576—about a million. Each combination of 20 bits could, for example, represent or be a code for any one of over a million different words. Or each combination could represent one number between 1 and 1,048,576. (Twenty-one bits, for example, could represent any one of 2^{21} words, or any number between 1 and 2,097,152—twice as many as 2^{20}). We would say that these words or numbers are represented in a 20- or 21-bit binary register.

Instead of representing whole words (using a 21-bit register as indicated above), one can use the $2^7 = 128$ combinations of a seven-bit register to represent any one item in a set including all of the following items: the twenty-six lowercase letters of the English alphabet; the twenty-six uppercase letters; the digits 0 through 9, and the punctuation, parentheses, and other common symbols such as those that exist on keyboards. Another of the combinations may be used to indicate an instruction, such as to begin a paragraph, and another may instruct either to begin or end a title or heading. Such a coding is precisely what is used to represent text in a binary digital transmittable form, in what is called the American Standard Code for Information Interchange (ASCII). This system has been used to enter, store, and display text in computers and (until 2008[3]) transmit text over the World Wide Web (the Internet).

THE STORAGE AND COPYING OF CLASSICAL INFORMATION

The storage of all information is physical. That is, information is marked or contained in a tangible, physical way in substances or devices. There was a time when one would have to hand copy a letter of some importance to remember what he or she had written or to send information to more than one recipient. Then came the printing press, which allowed multiple copies to be easily made—by some accounts this was the most important invention in human history. In modern times, we store or copy information as photocopies, audio or video recordings, photographs, e-books, or digital files. Information can be copied and transferred faithfully, especially if stored in digital form. And copying has made our lives easier in many ways. But then, copying has also caused a problem, as when copies are made in violation of copyrights.

CLASSICAL ENCRYPTION, AND THE CODE-BREAKING MACHINES OF WWII

The need for secrecy and the need to discover another's secrets has spurred an evolving "arms race" in the invention of coding methods and the corresponding invention of the means for code breaking (cryptanalysis) throughout history. Simple codes have been used since ancient times, including the substitution of one letter for another in any particular language. But in the common use of language there is a frequent repetition of certain combinations of letters (such as "th") or the frequent use of certain words (such as "the" and "and") that allow the codes to be broken with a relatively small amount of effort. Such code-breaking problems are often presented in the form of puzzles opposite the comic pages in our newspapers.

I introduce now, for illustration, a common, much more effective type of code that is not so easily broken, and one with a history, *the one-time pad.*

Suppose that Alice has a message that she wants to transmit in secret to Bob, her stockbroker. She wants to tell him to sell IBM stock. She can send a message in the following secure manner. (To simplify, we illustrate by coding to transmit only the letter "S," perhaps to indicate "sell.")

A. Alice expresses the letter *S* (the plaintext) as indicated below in ASCII form as a series of 1s and 0s. (Anyone knowledgeable would recognize this as ASCII and would be able to read the message.)

B. Alice adds to our ASCII message another arbitrary, randomly generated string of 1s and 0s (the key), known only to her as the sender and to Bob, the receiver.

C. Using the simple rule that any two 1s or any two 0s add to zero, but a 1 and a 0 add to 1, we get a digital sum called the ciphertext. If someone intercepts this ciphertext and uses the ASCII conversion to decode it, they will get another symbol or letter or number, not "S." (For the random string of key digits that we have chosen, an ASCII conversion actually yields a percent sign "%.") Were the ciphertext intercepted and read using ASCII, it would make no sense whatsoever.

D. Bob, however, adds the key to the ciphertext, and *voilà!*

E. He has the plaintext message that Alice wished to send to him, and he can convert using the ASCII chart to reveal the letter "S."

A.	1	0	1	0	0	1	1	The Plaintext
+ B.	1	1	1	0	1	1	0	The Key
= C.	0	1	0	0	1	0	1	The Ciphertext
+ D.	1	1	1	0	1	1	0	The Key
= E.	1	0	1	0	0	1	1	The Plaintext

In the above example, Alice has sent information to Bob in a ciphered binary string of numbers that no one except Bob can decode. Because Bob has the key that Alice has provided separately, he can easily decode the ciphered information to get the plaintext, and then use the ASCII conversion chart to read Alice's intended message.

Because sending different messages using the same key would soon allow someone else to deduce what the key is, this method of coding can only be used once or a couple of times with each key. And, of course, the success of this method requires that Alice be able to provide Bob with the key without anyone else gaining access to it, for then, they too would be able to decipher the secret message. The problem then is with *key distribution*. One famous (or infamous) example of key distribution involved the ciphered transmission of the rotor settings (the key) for the Enigma machine used by Germany just preceding and during World War II.

Many of us are aware of the Enigma machine, a system of rotors that provided the key for coding and decoding. Some of us saw the movie *The Imitation Game*, which is about the British effort to break its codes. The rotors could be set to work together in different ways to change the key, and instructions for the rotor settings would be sent out separately or somehow along with the coded message so the intended recipient could set the key. Without knowing the settings of the rotors of the machine, the key could not be determined.

Germany used ever more sophisticated versions of these machines to successfully send coded messages before and throughout World War II. It was the Poles who as early as 1932 mounted the strongest effort to break these codes, sharing their information with the French and the British and building the first computing machine for this purpose (called Bombas). The machine used electromechanical switches called *relays* to sort through the possible solutions to Enigma.

After the German invasion of Poland in 1939, the British—particularly benefiting from the genius of Alan Turing—were able to build a much more advanced and powerful machine, called Bombes; it was seven feet high by seven feet wide, and it was able to simulate the workings of thirty Enigma machines all wired together. However, the Allies were more often able to break these codes, not through any deficiency in Enigma, but through lapses in the way that the Germans used the machines, where, for example, repeated messages could be spotted and examined for patterns.

As Gribbin describes it, Turing designed Bombes to take advantage of the lapses, and his efforts probably shortened the war by two years, or managed to keep Britain in the war at all. He cites one example in particular: In the summer of 1941, Britain was on the brink of starvation. Thanks to the code breakers and Bombes located at England's Bletchley Park, convoys from the United States to England went twenty-three days without a single sinking, being warded away from the German U-boats with information provided from Bletchley.

But the Germans built an even more powerful successor to Enigma, called Tunney. Beating Tunney would be labor intensive. Gribbin paraphrases Turing: To comb for lapses "would require 100 Britons working eight hours a day on desk calculators for 100 years."[4] It became apparent that a newer sort of machine would be needed to break the codes. The first prototype of this machine began operating in Britain in June 1943, once again using electrical relays. Its successor, another prototype, this time used nearly two thousand electronic tube switches (somewhat like the radio tubes that preceded semiconductor transistor switches, to be described shortly). It was put into operation at Bletchley Park in 1944. Called Colossus, it filled an entire room. This was the first electronic computer, but it was programmable only in a limited sense.

MODERN CLASSICAL COMPUTERS (FAMILIAR TO MOST OF US)

Rather than tubes, modern computers use semiconductor electronic switches, transistor bits that are caused to carry either a small electrical current or no current, to indicate either a 1 or a 0, respectively. Sixty-five years ago, these switches were the size of a dime. Since that time, manufacturing technology has steadily advanced to reduce their size and integrate multiples of them and their associated electrical circuitry on wafers of silicon as single chips. In 1965, Gordon Moore, cofounder of Intel and Fairchild Semiconductor, noted that the number of components being manufactured in such an integrated circuit was doubling every eighteen months (mainly by reducing the size of the transistors and other circuit elements within the chip). This doubling rate has become known as Moore's law. It continues to describe progress to this day. (But the sizes of classical circuit elements have now been reduced to just one hundred times the size of atoms, quantum entities that loom as a fundamental limit to how small one can go. So, in ten to twenty years, Moore's law will probably no longer hold. Improvement in the capacity of classical computers through the reduction of element size will probably come to an end, and continued progress will require operation in a quantum realm. We'll soon get on to discussing quantum computers, but first let's examine what happens in classical computation.)

The information that is stored in classical electronic circuits is measured in *bytes*, where each byte represents eight bits of information. Today, each everyday desktop or laptop computer may store hundreds of billions of bytes, that is, hundreds of *gigabytes* of information.

Note that ten million bytes (that is, ten megabytes = ten thousand kilobytes = 10,000 KB) may be needed to store the information contained in an average book or photograph, or a sound recording. So the modern computer is capable of storing tens of thousands of these items.

But the computer does not just store information. It *processes* the information according to instructions that it has been given. The computer *processor* manipulates information using transistor switches that are "wired" in gates to perform logical operations. For example, the XOR gate (= Exclusive OR gate) will operate to produce a 0 if two input pieces of information are either both 1 or both 0. Otherwise, it will produce a 1. (These are the very operations that Bob used to combine the key with the ciphertext in our description of the "one-time pad" in the indented illustration A through E above.)

> The ten basic logic gates (all that are required to perform any logical operation or calculation) may be constructed as simple electrical circuits using combinations of transistors. For example, the XOR gate uses eight transistors appropriately wired together to take two input signals and yield one output signal according to whether the inputs are similar or not, as noted above. The input signals may come from other transistors, where the input "1" comes from a transistor which is "on" (carrying an electrical current in one direction), and the input "0" comes from a transistor which is "off," that is, carrying zero or little electrical current in that same direction. The transistor currents are turned on or off depending on electrical inputs brought to them from outside sources or still other transistors.

On another level, the computer may search and decide: "If *x*," "but not *y*," "then *z*," where *x* may mean "to have an appointment" and *y* may mean "to have a car" while *z* may mean "then take a bus."

Part of a processor's speed is determined by its *clock*, which puts out a regular electrical signal that coordinates the action of the digital circuits containing the transistors. The more frequent the clock signal, the faster the computer. But the clock signals must be spaced apart enough in time (be of low-enough frequency) that the transistors can complete their switching (of the 1s and 0s) between signals.

> The first general-purpose electronic computer, the ENIAC, was designed to produce artillery firing tables for the US Army. But it was soon diverted under the influence of the mathematician John von Neumann (then working at Los Alamos on the Manhattan Project) to run calculations on the feasibility of producing a hydrogen bomb. ENIAC's existence was announced to the public in 1946. It was 8 × 3 × 100 feet in size, contained 17,468 vacuum-tube electronic switches (bits), and ran with a 100-thousand-cycle-per-second clock (=100 kilohertz = 100 kHz). Each instruction took twenty clock signals to process, and so it had an instruction processing rate at 5 kHz.
>
> For comparison, note that my Mac mini desktop computer (which is four years old at the time of this writing) has a 2 gigabyte memory (2 billion bytes, 16 billion bits, a million times the bits in ENIAC) and runs with a clock speed of 2.4 billion clock signals per second (2.4 GHz), that is, 24,000 times faster than ENIAC.

A CLASSICALLY UNBREAKABLE CODE

As powerful as modern computers are, there are many things that they still cannot do, and that includes the breaking of certain codes. For one modern approach to coding, called *public key encryption*, the "public key" that is used to code a message cannot be used to decode the message. The reading (or breaking) of the coded message requires a related "private key." So, for example, Alice, who is to receive messages from Bob, may send to Bob as his public key a large number created by multiplying two large *prime* numbers together.

Prime numbers are numbers that have only two factors: the number 1 and the prime number itself. Said another way, prime numbers can only be divided (without generating fractions) by 1 and the prime number itself. Prime numbers include 2, 3, 5, 7, 11, and so on.

The encryption process in this case is not the simple addition rule noted in "C" of our boxed inset a bit earlier, and the public key can only be used to encode, not to decipher. So, Alice doesn't care if this public key is intercepted, because anyone having this number still will not be able to decode whatever Bob codes with it. Bob uses the public key to send a coded message to Alice. Alice uses her private key to decode the message that Bob sends. (Her private key is one of the prime numbers used to form the public key in the first place.) Alice never tells anyone what her private key is. And she doesn't have to send her private key anywhere, so no one can intercept or observe it.

Encryption using public keys formed of multiples of very large prime numbers is used by banks and the military to protect their transmissions. Such encryptions are also used to protect the information that you send when you enter your credit card information to purchase online.)

Decoding the message or information protected using the private key requires either knowing or finding one of the two prime numbers. (Having one prime and knowing the public key product allows an easy calculation of the second prime. The first prime is just divided into the number of the public key.) So, someone trying to crack the code must figure out what the primes are. Whereas, for example, finding the primes

3 and 5 that are multiplied to make 15 requires only a little trial and error, finding the primes of a number 400 digits long (such as may be used for the protection of credit card information) would likely take billions of years, even using the best of modern (classical) computers. And newer computers to come will not significantly alter the fact that, as a practical matter, this factoring can't be done in any reasonable length of time using classical computers.

QUANTUM COMPUTING

By contrast, a quantum computer having a processor of only 50 quantum bits would be able to run through the necessary code-breaking *algorithm* to find the private key in a few minutes.

> An algorithm is a self-contained set of logical or mathematical operations, such as "if *a*, then *b*" or "add *x* and *y*." By carrying out a set of these operations, a computer can make judgments or perform calculations based on input information.

No wonder, then, that efforts to build quantum computers are proceeding in a race funded by governments, big business, and the military. (As just one example of this, note that documents provided by Edward Snowden show that the US National Security Agency is funding a nearly $80 million program to develop a quantum computer "capable of breaking vulnerable codes."[5])

Because it is a quantum device, the calculations of a quantum computer ultimately involve probabilities, and so the result of a single computation may not be totally accurate. For the factoring task discussed above, this is not a problem, since the computer can always check its answers simply by multiplying the factors that it gets to see if their product matches that of the public key input. But, more generally, multiple calculations are run to provide a set of answers that will center around and define the correct answer. One can ask the computer to run the calculation as many times as is necessary to get an answer to a desired high level of certainty. This may extend the required computing time on complicated problems from seconds to perhaps a couple of hours, but that is still a very short time compared to the billions of years that may be needed to solve some problems or crack some codes using classical computers.

There are other intractable problems and classes of applications that may be addressed using quantum computers. These include faster search engines (like Google), rapid face and speech recognition (e.g., to identify one sought person—perhaps a terrorist—whose face appears in a photograph or video of a crowd of people), the simulation of all manner of physical and chemical processes, the design of new drugs and exotic materials, and efficient routing of people or the distribution of product. (Routing a salesperson efficiently to 14 different locations so that they don't incur unnecessary travel might take 100 seconds of classical computer time. But routing that same person to 22 destinations would require some 1,600 years using the best of today's computers.) Certain kinds of sorting problems are amenable to quantum computation. The example often used, just for illustration, is the reverse phone book, where one sorts to find the owner of a particular phone number. A classical computer can easily do this by performing, on average, perhaps a million operations (for a city of two million people). A quantum computer may do it with just a thousand operations (the square root of a million).

The *quantum bits* at the heart of the quantum computer behave much differently than classical bits.

> Quantum bits are commonly referred to as *qubits*. The term was coined in 1992 during a discussion between Professor Benjamin Schumacher and William Wootters, then of Williams College, initially as a pun in relation to the biblical measurement of length, the *cubit*. (For instance, Noah's ark was so many cubits long, so many wide, and so many high.)

What distinguishes qubits is that they can be in a *superposition* of both of two states *simultaneously* or be in either of the two states, based on probabilities. This is not a simple statement. Qubits can be formed and behave in this manner because of the fundamental quantum makeup of our world.

To illustrate how qubits might be created physically, consider again the single electron in the hydrogen atom.

The overall wavefunction for the electron is a superposition of all of the possible electron's states, represented by all of their wavefunctions and all of their associated probability clouds (like those shown in Fig. 3.8), all superimposed and centered around the nucleus of the atom. In nature, things tend toward lowest energy, and so the probability of finding the electron (upon measurement) in its lowest energy "ground" state is greater than for the higher energy states. This probability information is also contained in the overall wavefunction. The wavefunctions may evolve and change with time. But until an observation (measurement) is actually made, the electron remains in a superposition of states and clouds according to the probabilities of the overall wavefunction.

The act of observation (measurement) selects one of the allowed states. But as soon as the observation is over, probabilities take over again, and another, different, overall wavefunction describes the electron and its new evolution in time.

Hypothetically, for illustration, suppose that a qubit can be created in a hydrogen atom by controlling an electron to switch occupancy between just two particular states, one representing a "0" and one representing a "1." (The switching might be caused by shining laser light of just the right frequency on the atom, either inducing a transition of the electron from the higher to the lower energy state or, through the absorption of a single photon, pumping the electron up from the lower energy state to the higher energy state.) In this two-state arrangement, the overall state could be described as a combination of the two states, with the overall wavefunction containing the information on the probability of finding the electron in one of these states or the other.

As you will soon see, there are many types of particles and properties that are being developed to produce a binary combination of qubit states in somewhat this manner. In such binary physical systems, a combination of two states (and two-state wavefunctions) describes each qubit.

Two qubits in proximity can affect each other if their wavefunctions significantly overlap. In the right circumstances, each qubit can actually be linked by *entanglement* with the other qubit, so that together they are described by a new overall evolving wavefunction that treats them as a single entity. And this entanglement, once established, under the right conditions can be made to persist even if the two physical qubit objects then move far away from each other. So making an observation or measurement on one of the qubits will not only choose its state, but it will also immediately determine the state of its entangled partner, no matter how far away it is. (As you may recall, this ability of particles to stay entangled over long distances of separation is what Einstein called "spooky action at a distance.")

The pair of qubits can be in any of the 2^2 = four combinations of their states or in a *superposition of all four combinations simultaneously*. (These four two-qubit combinations can represent the four binary digit combinations: "0, 0," "0, 1," "1, 0," and "1, 1.") As many as n qubits can all be further linked, in which case 2^n possible values (or numbers) can be *simultaneously* digitally represented and stored by n different strings, each of n 0s and 1s. (This is distinct from n classical bits, which can represent and store *only one* of the 2^n values at a time.) And remember, if n = 20, then 2^{20} = 1,048,576. That's a lot of simultaneously stored numbers for just twenty qubits!

So, why can't we link and entangle classical bits together in superposition for simultaneous operation in the same sort of way? We can't, mainly, because classical bits are necessarily spaced too far apart: their wavefunctions have no significant overlap. They are deliberately spaced apart so that the operation of one bit doesn't interfere and cause problems with the operation of its neighbor, like the shorting of current from one bit to another without traveling through the connecting circuitry.

Until recently (2015) the transistors (classical bits) and the deposited copper lines (acting as miniscule wires) that are used to electrically control transistor switching in integrated circuits have been thousands of atomic diameters across and separated by comparable distances. The *diameter* of an atom can be thought of as the distance over which the wavefunction of its lowest-energy (most likely occupied) state drops to essentially zero (where its probability cloud disappears into blackness). So, if the transistors are separated by thousands of atomic diameters and the wavefunctions of the atoms that they are made of essentially disappear over a distance of a few atomic diameters, there is little wavefunction overlap and cooperative behavior.

But, as pointed out earlier, transistors and circuit elements are now being made in sizes just hundreds of atoms across, and continued classical miniaturization according to Moore's law will soon (over the next ten to twenty years) reach a quantum limit as element dimensions approach the sizes of atoms and their wavefunctions begin to substantially overlap. Like it or not, then, continued miniaturization will involve inadvertent quantum connections. Work already started (to develop quantum bits) may be necessary for further progress in miniaturization, and to otherwise increase the data storage capacity and computing power of computer elements.

Quantum qubit properties also allow the formation of new types of logic gates and new and simpler algorithms for new and still faster quantum processing. To get a sense of this, note that one quantum-based algorithm to factor a number n uses the mathematical expression ($n^2 \log n$), which is far less complex and requires many fewer operations than the classically based factoring algorithm, which uses ($\exp[n^{1/3}\{\log n\}^{2/3}]$).[6]

One particular logic gate, the CNOT gate (not possible with classical bits), operates by flipping the state of a target qubit only if (for example) a second control qubit is in a 1 (as opposed to a 0) state. What makes this process special in another way is that the two qubits can become *entangled* in the CNOT operation, linked together to represent any one of their four so-called Bell states,[7] regardless of how far the qubits may later become physically separated from each other.

As noted earlier, the power of quantum information storage and processing (e.g., the running of multiple programs simultaneously using the same qubits) derives from the superposition of states within a qubit and the entanglement of two or more qubits. But remember, any outside contact, observation, or measurement on entangled particles or qubits introduces decoherence and breaks the entangled state, as described near the end of Chapter 6. So the quantum computer must be constructed so that it chugs away, doing multiple tasks in superposition without observation or measurement until an answer is found and is then read (observed).

Creating the physical qubit elements, linking them by entanglement, and maintaining the entangled state is not an easy matter. The qubits must be isolated from their surroundings so that outside influences (like the thermal oscillations of atoms normally present in solid materials) will not decohere the quantum entangled state. But they may not be so isolated one from another that they can't be linked by entanglement, and they must still be sufficiently accessible so that their initial states can be manipulated and their final states can be observed and read. Gribbin lists five requirements for qubits, summarized as follows[8]:

1. They must be well characterized and scalable to production in large numbers.
2. One must be able to initialize their states to set up the computation.
3. The "decoherence time," that is, the duration time of their entanglement, should be long enough to accommodate a million gated operations.
4. They must incorporate reversible gates and be able to work around errors.

5. They must be able to repeat their computations many times to be able to provide reliable readouts.

And Gribbin goes on to describe a half dozen approaches being explored for the construction of qubits meeting these criteria, tracing the progress of their development and their use in computers until 2014.[9] Note: Here I use terms and concepts that may become more understandable after reading about the foundations of chemistry and materials science in Part Four and the nature of various superconductor and semiconductor devices in Part Five of this book. But this is a preview of coming applications, and so I leap ahead.

The following approaches are being explored toward quantum computing, each with particulars and advantages and disadvantages, as described in Appendix C. As you will come to understand, quantum computing is still an application in the infancy of its development, but one that holds great promise.

1. Ion traps, where it all began;
2. Nuclear magnetic resonance, so far used to factor the number 143 into its constituent prime numbers;
3. Quantum dots, promising devices that operate at ambient (room) temperature, and projecting 20-qubit circuits in five to ten years;
4. Isotopic nuclear spin, with long coherence time;
5. Quantum photonics, using polarized states of particles of light, so far to factor the number 15 into its primes, but also relevant to quantum encryption and quantum teleportation, and, perhaps the most advanced of approaches,
6. Superconducting Quantum Interference Devices (SQUIDS), which so far include the development of a 1,000-qubit processor (announced in June 2015). (However, note that the actual quantum computing capabilities of this processor may still need to be demonstrated.)

QUANTUM TELEPORTATION (IT'S NOT WHAT YOU THINK IT IS.)

Quantum teleportation of a very basic form has already been demonstrated, and I'll soon get on to describing that. But teleportation as envisaged in *Star Trek* ("Beam me up, Scotty"), although possible in principle, seems highly, highly, highly unlikely. What makes it possible at all is a quantum process, because the constituents of our bodies are all quantum in nature, comprised of atoms and molecules, as described briefly below.

A Refresher on the Quantum Atom

As you have already come to realize from Schrödinger's description of the hydrogen atom, the states of the electron in an atom are diffuse entities, with the position of the electron in the atom described in terms of clouds of probability, as shown for the hydrogen atom in Figure 3.8. Which of the states are occupied by electrons is also based on probabilities. Like most things in nature, every atom seeks a lowest-energy state, a ground state where all of the electrons occupy the lowest energy levels allowed to them, but there is always some probability that they will be in some higher energy state. The same description of states and probabilities applies to molecules, atoms in chemical combination.

While teleportation in the *Star Trek* sense would require the teleportation of all of the atomic and molecular constituents of the human body, the teleportation of just one atom should suffice to demonstrate the possibility of this actually happening in principle. We'll consider the teleportation of the simplest of atoms, hydrogen. But let's first examine the possibility of teleportation using classical means.

Is It Possible to Achieve ***Classical Teleportation***?

The answer is no. We can't teleport an actual physical object by any known classical method. But we can teleport classical information; we do it all the time, using fax machines. Why not then measure or observe the properties of an object and then fax that information (using phone lines and satellite transmission at the speed of light) so that a copy of the physical object can be constructed far away based on the information that is sent? Surely we should be able to do this quickly on one atom to demonstrate the principle.

Well, there is a problem even with this. The very act of measurement or observation changes the state of the original atom that we want to send. It is the information on this changed state that we then have and then send. Information on the original state of the atom, with all of its probabilities, is not preserved and not available to us.

In short, we have teleported a description of *an* atom—but not *the* atom that existed originally. A human constructed of atoms and molecules teleported in this way (though it might have the same numbers of atoms in the same places as in the original) would in fact not be the same person we were intending to teleport.

"But," you may say, "we have teleported a body!" Sure, but instead of Scotty beaming up a particular person (say, John) by monitoring the state of every atom in John's body, we've caused every atom in him to change, and we've sent the information on the changed person on to Scotty. He constructs the changed person at the other end. It's not the original John that we've teleported.

So How Is ***Quantum Teleportation*** Different?

Well, quantum teleportation does not involve the direct *measurement or observation* of the properties of the particle that we wish to teleport, and so no decoherence or change takes place, at least not initially. Instead, the quantum properties of the particle to be teleported are linked in an entangling operation to the properties of another particle that is already entangled with a third, perhaps faraway, particle. A final operation (guided by the separate transmission of classical information describing how the first operation was performed) then causes the faraway particle, through its entanglement, to take on the properties of the first particle. The first particle *is* altered in the process (but not before its original properties are linked to those of the faraway particle). The properties of the original particle have thus effectively been teleported to the faraway location. (It is important to note that it is the *properties* of the original particle that have been teleported, *to another already-existing, faraway particle*. The original particle itself has not been teleported, has not been moved, but it has had its properties altered in the process.

Such teleportation has already been demonstrated for the case where all of the particles are photons. We examine now this simple case. Then we'll move on to consider the teleportation of ions, atoms, and molecules—the stuff that humans are made of.

How the Quantum Teleportation of Photons Works

Note that all photons can be used as binary qubits, that is, each can be made to exist in either a vertical or a horizontal polarized state or in a superposition of these two states. (For a definition and physical description of polarization, refer to Appendix A and Figure A.1(c).) We might assign a 1 to represent the vertical state and a 0 to represent the horizontal state. Then we're back in the realm of quantum computers and quantum information processing, all of which is formally described as a part of quantum information theory. But so is teleportation.

In one application of the theory, particles having binary states can be teleported through a well-defined process. We illustrate with our photons. In the first step, an entangled pair of photons is created, what is called an *ebit*, and each still-entangled photon is then sent off to be shared: say particle E1 to Alice and particle E2 to far away Bob.

Sometime much later, in principle, maybe months later, Alice may have a photon that she wishes to teleport to Bob. Let's call it T. This photon is also a qubit polarized in either of two perpendicular directions or in some combination of these two polarized states.

Through the performance of what is called a Bell measurement, E1 becomes entangled with T. This measurement does not decohere or change the state of T, because there is no observation of its polarization; it asks only if the polarizations of the two photons are the same or not, and it entangles them in the process of finding out.

Alice does not know the polarization of T, but she does know the parameters of the Bell measurement that has been performed. She sends Bob a description of the measurement. She does this using classical methods, perhaps using radio or Internet communications that can take place at the speed of light (including the operation of servers, etc., it might take a bit longer).

Bob uses the information that Alice sends to effectively reverse the effects of the Bell measurement and leave E2 in exactly the state that T was in originally. T is left at the end of the process in a different state than it was in originally, but because E2 has T's original properties, the original T has been effectively teleported from Alice to Bob.

"But," you might say, "Why didn't Alice just send T directly to Bob? T is a photon, so it travels at the speed of light, in fact *is* light. Then Bob would have the original T at the speed of light without all of the intermediate finagling."

Alice could, in fact, do that. But we have three purposes in describing photon teleportation. First, something like the teleporting of photons can be used to send encrypted messages in ways that are totally foolproof, as will be described next. Second, quantum computers could use teleportation to connect remotely located internal quantum components, or to communicate directly and securely computer to computer, bypassing the classical stages of supplying input and reading output (essentially eliminating human or classical devices as "middle men"). And, finally, particularly as the teleportation of photons has actually already been tested, we want to use this example with photons as a basis for considering the teleportation of objects with mass that can't travel at the speed of light.

Tests of Photon Teleportation

In 2004, a group under Anton Zelliger successfully teleported photons 600 meters (about 600 yards) through optical fibers from one side of the Danube River to the other side.[10] Gribbin also describes two later significant tests, both of which occurred in 2012.[11] In the first test, a large group of Chinese researchers managed to teleport "a quantum state" through open air over a distance of 97 kilometers (about 60 miles). In the second test, a team from four European countries teleported "the properties of a photon" some 85 miles at an altitude of nearly 8,000 feet (where the interfering air is thinner) between stations in the Canary Islands, on Las Palmas and Tenerife. If transmission is easier at altitude, why not then transmit first to satellites as relays and then back down? The Chinese envisage eventually creating a secure quantum network constructed in this manner, requiring multitudes of entangled photons. As of 2012, toward this sort of goal, Chinese researchers have been able to produce entangled groups of four photons at the rate of several thousand per second.

Entanglement is the key to teleportation. But for the teleportation of bodies, we'd need to entangle particles of matter, not just photons. And we'll further require an entanglement of the complete set of wavefunctions describing all of the states of each atom or molecule, not just a binary pair of states. Some progress has been made.

The Entanglement of Ions and Teleportation of Electrons

In *Computing Quantum Cats*, Gribbin describes work at the University of Michigan in which two ions about three feet apart are entangled with each other through the entanglement of their emitted photons.[12] And an article in 2014 reports an achievement at the Delft Institute of Technology in the Netherlands, of entanglement between single electrons trapped in supercooled diamonds placed thirty-three feet apart.[13]

These are major accomplishments, but we clearly have a long way to go, to teleport even single atoms over very long distances. Ultimately to completely teleport a human being requires that a staggering number of atoms be teleported: the order of ten billion billion billion is a good approximation. We might conclude that the teleportation of humans à la *Star Trek*, although possible in principle, is exceedingly unlikely.

Remember, too, that unlike in *Star Trek*, we are only teleporting properties to atoms that already exist at a faraway destination. We aren't transporting the atoms themselves. And the atoms that stay behind have changed properties. The faraway atoms now resemble these earlier atoms before they had changed. (This, by the way, is what distinguishes teleportation from cloning. Had the earlier atoms retained their properties, then we would have two identical beings, one of them cloned from the other. But this can't happen in a quantum system. There is a "no cloning" theory, derived from quantum information theory, that shows that copying and cloning are not possible. The original state is always changed, and so one is always left with teleportation. On the other hand, classically one can copy and teleport, but not clone. Interesting!)

But here's food for thought: Orzel notes[14] that some scientists (Roger Penrose, for example, in his book *The Emperor's New Mind*) suggest that consciousness is essentially a quantum phenomenon. If so, Orzel points out, teleportation might be used to transport the contents of our minds. (I would add: if that's the case, do we really need our bodies? What then of mortality?)

ABSOLUTELY SECURE QUANTUM ENCRYPTION

The very ability to copy is what makes classically transmitted information vulnerable to eavesdropping and surveillance, often without the sender and receiver even knowing about it. Even the public-key-encryption methods described earlier are susceptible in principle to decoding because the message sent can be copied. For example, a message can be copied into a bank of computers that operate in parallel toward breaking a code. Classical computers may lack sufficient power to do this with modern public-key encryptions, but soon quantum computers will succeed, as described earlier.

In contrast, quantum information simply cannot be copied by either classical or quantum means. To read or make a copy would physically involve an observation, a measurement. And, as we have learned, an essential feature of our quantum world is that any measurement or observation causes a decoherence of the wavefunction of an object, destroying much of the initial information that the wavefunction contained. What is measured is only a transformed part of the "original," and the "original" exists no more.

This inability to copy quantum information *is at the heart of an ability to encode* information in ways that cannot be deciphered by anyone or anything other than the intended receiver, and further cannot even be tampered with without the sender and receiver knowing of it. The principal use of this ability is in what is called *Quantum Key Distribution (QKD)*, described as follows.

Quantum Key Distribution—How It Works

Remember our first example of classical cryptography: Alice sent an encrypted message to Bob using a method called the one-time pad. Bob was able to decipher the message using a key, a string of 1s and 0s sent separately from Alice to Bob beforehand. As long as Bob has the key, he can decode even very long and complicated messages. But sending or distributing the key, especially to a number of designated receivers, can get expensive, and if the key were to fall into the wrong hands, unintended other persons could decipher the messages without the sender or the intended receivers ever knowing about it. One might say that securely sending The Key, is key.

Quantum Key Distribution (QKD) offers foolproof methods for sending the key. One QKD protocol, called *BB84* (after Charles Bennett of IBM and Gilles Brassard of the University of Montreal, who together invented it in 1984) uses qubits to transmit a key in a coded way. Another QKD protocol, called *E91* (after Artur Ekert, who suggested it in 1991), involves operations on a set of polarized entangled photons, in a manner somewhat similar to that described earlier for photon teleportation. For either protocol, any attempt to intercept the entangled information being sent destroys the entangled state in a way that is detectable by both sender and receiver and makes the key unusable. There is no known method (even in principle) for cracking an encrypted communication made using a one-time quantum key established by either of these methods, and this will hold true even when there are more powerful classical or quantum computers.

Quantum Key Transmission

Secure BB84 keys have been exchanged through optical fiber lengths of 12 miles at rates of a million bits per second, and of 60 miles at rates of ten thousand bits per second. (Typical optical fiber communications are 1,000 to 10,000 times faster. But remember, it is only the key that needs to be transmitted in this secure quantum manner; the rest of the message can be encoded using the key and sent at classical speeds.) The longest quantum key transmission has been achieved with optical fibers through a length of about 100 miles. And the same European group that achieved a teleportation of 80 miles through high-altitude air (between stations at two of the Canary Islands) has similarly managed secure QKD transmission through air over the same distance using both the BB84 and E91 protocols. Since the air is thinner yet at the altitude of satellites, this suggests the possibility of secure longer-distance transmission through satellites as relays.

Quantum Key Distribution Networks

These have been set up in the United States at the Defense Research Projects Agency (DARPA), in Vienna, in Switzerland, and in Tokyo.

Commercial Quantum Key Distribution

As of December 2015, at least four companies are listed as manufacturing quantum cryptography systems: MagiQ Technologies, Inc., of Boston; ID Quantique of Geneva; QuintessenceLabs of Canberra; and SeQureNet of Paris.[15] As practical quantum computers that are capable of factoring large products of prime numbers become available (so that classical public-key encryption is no longer safe), these companies should experience an increased demand for their quantum key-encryption services.

CHAPTER 8 SUMMARY

The takeaway points from this chapter are: (1) that quantum computers offer the possibility of computation *for some applications* that can be millions of times faster than classical computers; (2) that teleportation is not possible by classical means; (3) that teleportation is possible *in principle* using quantum methods, but, as a practical matter, for much more than single particles it becomes extremely difficult—and for humans highly, highly, highly unlikely; (4) that the teleportation that would result is a teleportation of properties, not particles; (5) that quantum computers are a threat to breaking some of the best of our codes; but (6) quantum encryption offers codes that cannot be broken, and cannot even be tampered with (without the sender and receiver knowing that someone has interfered).

This chapter involved information theory as applied to encryption and teleportation using photons and matter on a quantum atomic scale. In Part Three, Chapter 9, we'll see how the potential loss of information in the evaporation of black holes threatened the theory of quantum mechanics. This as just one small part of our consideration in that chapter of the quantum particles of nature and the quantum's role in the expansion of our universe, from the big bang to the galaxies.

Part Three

OUR WORLD OF RELATIVITY AND THE QUANTUM, FROM THE BIG BANG TO THE GALAXIES

In Part Four of this book I describe how quantum mechanics provides an understanding of the quantum nature of the atom, of chemistry, and of the physics of materials, the intellectual elements from which wonderful modern inventions have been created. Many of these are described in Part Five.

But before we turn to these more practical aspects of the theory, I want to give you a sense of other exciting aspects of our world, what we know of the smallest things that existed from the beginning of time to the present day, and of the galaxies, and of the overall structure of the universe of which they are a part. This is a world of relativity and the quantum—a fascinating world.

In explaining this world, it is useful to use analogies, both to provide insight and to connect the parts of what is presented. I am assuming the role of your tour guide in a time-travel visit of our universe. I have already been to many of the fascinating places that we will visit, often as a visitor in other tours. I have usually been escorted by local guides (authors, through their books). Sometimes these are people of authority, but often they are people skilled at providing me the information from authoritative sources in a way that I can more easily understand. (Some of these writers may be thought of in analogy to the guides at galleries or architectural wonders, or the docents who may lead us through museums.)

As your overall tour guide, I have selected those topics that I have found to be particularly interesting, and I have arranged your tour in a way that logically sequences and connects them within an overall journey that takes us backwards from the present to the beginnings of time. Some of the places at which we will stop (topics that we will cover) are included in the title of Chapter 9. I have chosen a good set of guides for us on each topic, and, because these guides often have somewhat different interpretations of their subject matter, they may emphasize one thing or another.

While most of what will be described is based on hard evidence and well tested and proven theory, in some cases what the local guides and drivers present may be conjecture or even opinion. In these places, I will point this out and qualify what is presented. And I will, in most cases, introduce you to the guides and drivers (by referring you to my sources) so that you may explore further with them directly. Because we make brief stops in the various topic areas, and because they are many and by my design flow one to the next, I have not chosen to break their descriptions into separate chapters. Rather, all topics are included in this single Chapter 9, which comprises all of Part Three.

While this chapter is nominally structured as a tour of specific topics, the topics are tied together to leave you with an understanding of two pillars of modern physics that describe our world: the big bang model, which traces the evolution of our universe; and quantum mechanics, which applies generally in what is called the Standard Model and describes the fundamental particles that either transmit the forces of nature or (for some of them) lie within the atomic building blocks that make up everything that we see around us.

I welcome you now to a trip of great discovery and beauty, with concepts and observations that may challenge your view of the world around and within us.

Chapter 9

GALAXIES, BLACK HOLES, GRAVITY WAVES, MATTER, THE FORCES OF NATURE, THE HIGGS BOSON, DARK MATTER, DARK ENERGY, AND STRING THEORY

This chapter is divided into four sections:

I. Obtaining a Map of the Universe (Basically an Introduction to Understanding the Universe in Which Our Trip Takes Place)
II. Sightseeing (A Quick Tour through the Evolving Universe, from the Big Bang and the Very Small to the Present and the Very Large, with Stops Afterward to Explore Particular Sights, including Those Topics in the Title to This Chapter)
III. Key Aspects of the Big Bang Model
IV. Approaching the Big Bang (Creating the Conditions of the Hot "Quark Soup" just after the Big Bang, to Explore the Fundamental "Building Block" Particles of Nature and the Particles That Convey Nature's Forces)

Section I

Obtaining a Map of the Universe

(Basically an Introduction to Understanding the Universe in Which Our Trip Takes Place)

A. Space, Time, and Relativity
B. Models of the Universe
C. The Big Bang Model

A. SPACE, TIME, AND RELATIVITY

Orientation

We begin now what is probably the most difficult part of our trip: Subsection A of Section I. In this section we are going to briefly leave what most of us conceive of as our universe and enter another one, our actual world, that is forged in the realities of relativity and the quantum. I don't expect you to fully grasp what is presented here, but I need you to just "hang in" and listen, for this is the universe that we live in and what the big bang model describes. (I start with this orientation because I've found that tourists are often otherwise confused. They need this information up front.)

Special Relativity

We'll start slowly by describing some aspects of relativity. First note that observers who are moving relative to each other will get different results when making measurements in space and time. Such differences have actually been observed, as described in the box below.

Objects moving relative to us appear to us to shrink in the direction of their motion, and clocks moving with them are found by us to tick at a slower speed. We normally just don't notice these things, because at the low velocities that we deal with the changes in the objects and the clock are imperceptible. But we do see related effects in high-energy physics accelerators, where particle speeds approach the speed of light. And, with the precise instruments available to us in recent years, we can actually measure these effects even at the much lower speeds of airplanes.

Part of Einstein's genius was to recognize, as described in his theory of special relativity, that physics is the same for every observer and that we can use one frame of reference and relativity theory to correctly predict what observers in the other frames of reference will see. That frame of reference is a four-dimensional combination of space and time called *space-time*. (Here time is in some sense a fourth dimension along with the three spatial dimensions familiar to us: up and down, forward and back, and sideways). The problem is that we find it very hard to visualize four dimensions, let alone a fourth dimension in time. (Should you wish to get more than this vague sense of space-time, I recommend the very readable book by Barry Parker on Einstein and relativity.[1])

General Relativity

Now, realize that the Newtonian classical physics (familiar to us) and Einstein's special relativity are both special cases of Einstein's general theory of relativity (henceforth "general relativity"). Special relativity describes relativity only as an approximation in the limit of small gravitational fields. And Newtonian physics is an approximation of special relativity that is accurate only in the limit of small relative velocities [and small gravitational fields]. But general relatively describes our world for all events whether in large or small gravitational fields and regardless of relative velocities. It predicts, for example, not only that clocks will tick more slowly for an observer when the clock is moving relative to that observer, but also that an observer will see the clock tick more rapidly if the clock is in a gravitational field that is reduced compared to the field at the observer. Here are just a couple of examples of what is observed.

The gravity of the earth diminishes as we move out from the earth's surface. On a hypothetical surrounding sphere twice the earth's diameter, the earth's gravity is 1/4 what it is at the earth's surface. On a sphere three times the earth's diameter, the earth's gravity is 1/9 what it is at the earth's surface, and so on.

Our global positioning system (GPS) works using 31 global positioning satellites that orbit 12,500 miles above our earth (on a hypothetical sphere of about four Earth diameters) and measure the time of flight of signals that they receive from our phones and other devices. From this collection of measurements, this system can normally indicate the position of our devices to within six inches. If corrections were not made for the effect of gravity, an accumulated error in the measurements over the course of a day might total six miles.

And the clock in an airplane, when taken to an altitude of, say, 30,000 feet, will actually tick faster (rather than slower) to an observer on the ground when compared to a clock at the earth's surface. That is because the lower gravity's tendency to make the clock run faster will, at that altitude, be greater than velocity's tendency to make it appear to run slower.

These are interesting effects to illustrate that relativity theory predicts and explains actual observations. Just examples. There is much more to it.

Now Comes the Overarching Concept

As with quantum mechanics, general relativity has successfully described nearly every aspect of our universe in which it has been tested. And general relativity's role in cosmology is profound. Space-time is *part of* our universe. It is influenced by, and influences, the matter and energy within it.

As the physicist and writer Bojowald describes it: "The form of space-time is determined by the matter it contains. Space-time is not a straight, flat, four-dimensional hypercube extending unchanged all the way to infinity. Like a piece of old rubber, it writhes under its own inner tensions into a curved structure. The inner structure of space-time is the gravitational force."[2] (According to Einstein, gravity is not a force, but it appears like one because of the curvature of space-time.[3] This is a really counterintuitive concept, shocking even to physicists when it was first introduced. More on this later. Despite the fact that they now recognize and use curvature as Einstein does in their explanation of gravity, many physicists and cosmologists still tend to refer to gravity as a "force.")

Later, referring to Einstein's accomplishment, Bojowald writes: "Promoting space-time from a mere stage, serving only to support the change of matter, to a physical object in the theory of relativity is a revolution." And "the role of space-time, now seen as a physical object, is often compared to a novel in which one of the characters is the book itself."[4] (Pretty heady stuff!)

This then is the relativistic framework of the universe that we explore. Its evolution has been shaped by relativity (and, as you will see, also the quantum), because that's just the way the universe is. "Theory" and "modeling" are only our tools for understanding what we observe.

Einstein assumed that our universe is roughly the same everywhere. This has come to be known as the *cosmological principle*. He also assumed that there is no center to the universe and that any place in it (on a grand scale) is pretty much comparable to any other. This is known as the *Copernican principle*.

The latter principle is named with reference to the observation and calculations by Nicolaus Copernicus in 1543 (which he only dared release on his deathbed), that the earth is not the center of the universe. The idea of an earth-centered universe was generally believed at the time and taught specifically by the Catholic Church, since that idea generally supported the biblical text. Galileo, sometimes called "the father of science," who observed the skies with his own telescopes, spoke and wrote in support of the Copernican idea. He was tried by the Roman inquisition in 1615, was forced to recant his views, and then was placed in house arrest for the remaining nine years of his life.)

Including these assumptions, Einstein found that general relativity predicted either an expanding or a contracting universe, unless he adjusted a "cosmological constant" that would allow him to get the result that he expected at the time—a static universe, neither expanding nor contracting.[5]

But the Russian mathematician and meteorologist Alexander Friedman and the Belgian Jesuit priest Georges Lemaître later used Einstein's equations and concluded that the universe began "as a tiny speck of astounding density . . . which swelled over the vastness of time to become the observable cosmos."[6] According to bestselling author Brian Greene (writing as professor of physics and mathematics at Columbia University), Einstein faulted Lemaître for "blindly following the mathematics and practicing the 'abominable physics' of accepting an obviously absurd conclusion."[7] It was only a couple of years before observation resolved the issue.

We board the bus to the planetarium. From here on, because it's easier for us to visualize and we'll lose little in meaning, we'll usually talk of our familiar space and time, rather than "space-time."

B. MODELS OF THE UNIVERSE

At the Planetarium—The Expansion of Our Universe

Here we first look at the stars and the galaxies, as was done historically. Then we'll examine the model of our universe that will guide our tour later on.

Evidence for Expansion of the Universe

Strong evidence lies in the discovery by astronomer Edwin Hubble in 1929 of a Doppler redshifting (see the next paragraphs below) of the wavelength of light arriving from the stars of distant galaxies. This was some five years after Hubble obtained the first clear view of nearby galaxies that had been seen before in earlier, smaller telescopes only as faint, fuzzy nebulae.[8] (Hubble used the 100-inch Hooker telescope located on Mount Wilson. Today, using radio telescopes, the Hubble telescope [named after Hubble, but placed in orbit around the earth long after he was gone] and many other types of instruments, we have evidence of some 100 billion galaxies in just that portion of the universe that we can observe. Though they vary greatly, each of these galaxies contains on average over 100 billion stars.)

Doppler shifts occur in waves of various kinds as they are emitted from moving objects. Speeders and baseball fans may know that radar guns use the Doppler effect. And some of us are familiar with the weatherman's observation of the movement of rain, snow, or hail using Doppler radar.

These devices work by sending out electromagnetic microwave pulses and observing the changed wavelength of their reflections. What we see from the stars is light that is emitted rather than reflected. But Doppler shifts for all electromagnetic waves occur in the same way that they occur for sound.

You may have noticed that the engine sound from a race car or the siren of a police car increase in pitch (frequency) when it moves toward you, then suddenly drops to a lower pitch as it passes and moves away. These changes in frequency are referred to as *Doppler shifts*.

Similarly, light emitted from an object moving toward us has an increased frequency (and decreased corresponding wavelength), while light emitted from an object moving away has a decreased frequency (and increased wavelength). Because red light is at the longer-wavelength end of the optical spectrum, this latter going-away shift is referred to as a *redshift*.

When Hubble used his telescope to look in any direction at distant galaxies, he found that the wavelength of the light gathered by the telescope is redshifted, indicating that the galaxies are moving away from us. (This would mean, for example, that the spectral lines shown in Figs. 2.8 and 2.9 would be shifted to the right.)

He also found that the more distant the light source (distance revealed by the intensity of light from Cepheid stars of known brightness), the more extreme is the redshift, indicating that those galaxies farthest away are moving away from us at even greater speeds. Though Hubble made his observations on a relatively few galaxies, what he concluded has been supported by all observations since then.

That these observations have been found to hold universally for faraway galaxies is now referred to as *Hubble's law*. The population of distant galaxies throughout space also looks pretty much the same, as far as we can see in any direction. What emerges from Hubble's observations is a picture of an expanding universe. Friedman and Lemaître were vindicated!

(I have been referring to "distant" galaxies moving away from each other in this expansion. For galaxies that are a sufficient distance apart, continued separation because of expansion dominates. But nearby galaxies can actually be moving toward each other faster than the local expansion of space pulls them apart. And so it is that our nearest-neighbor galaxy, Andromeda, is on a course to collide with our Milky Way in about four billion years.[9])

Now, so that you have the rest of the overarching background, realize that within the expanding-universe scenario there are a number of mathematically defined, impossible-to-visualize, possible "shapes," to the space-time of the universe, each having what is defined as "curvature."[10] Based on the shape and curvature, our universe might be viewed in different ways.

In his book *Big Bang, Black Holes, No Math* and in his course, Toback describes three scenarios, based on the density of matter and energy that can be directly seen in the universe.[11] Above a certain critical density, the expansion no longer continues and it contracts back to infinitesimal size in what he calls a "big crunch." (Matter attracts and causes contraction.) With less than the critical density, it expands forever. With just exactly the critical density, it expands, but at a decreasing, "decelerating" rate. Well, the total of mass and energy *that can be directly seen* is woefully short of the critical density, so we'd expect continued expansion.

But Toback describes two more components to our universe, "dark matter" and "dark energy" (neither of which are directly seen, both to be described in Section III (B), toward the end of this chapter). As it happens, when these two components are included, the density of all known matter and energy in the universe is exactly 100 percent of the critical density. (Isn't that interesting!) Dark energy, which contributes 72 percent of this density, is inferred from observations in the late 1990s of an accelerated expansion. (As opposed to matter, energy can produce a repulsion, which spurs expansion.) Our best understanding now is that, until about six billion years ago, after a very short initial period of very, very rapid expansion called "inflation," the expansion was slowly decelerating. But in the last six billion years, because of this dark energy, our universe has been expanding at a faster and faster rate. We call this recent period "acceleration."

But we're at 100 percent of critical density, and, as described by Greene in his very popular book *The Hidden Reality*, 100 percent of critical density makes the curvature of the universe "zero."[12] This zero curvature is referred to as providing a "flat" universe (not flat in any geometrical sense that we can visualize, though). Within this "flat" universe are two possibilities: either the universe is infinite in extent, or it somehow has an "edge," in analogy to the edge of the screen of a video game. (Think about how Ms. Pac-Man crosses the left edge of the screen and suddenly appears way over on the right edge.[13]) Greene goes on to say that there is nothing that distinguishes between these two possibilities, but that physicists and cosmologists tend to favor the infinite extent (with no sort of edge). He then makes the case that infinite extent implies (as distinct from "many worlds," which we described in Chapter 6) the existence of a "multiverse" of parallel universes, all identical to the one that we occupy, with identical occupiers and identical events transpiring in each. (This multiverse, and others, are beyond the scope of *Quantum Fuzz*, but [for those who may wish to explore further] I note that they are part of a set of topics included in Greene's already cited book *The Hidden Reality: Parallel Universes and the Deep Laws of the Cosmos*, Reference AA.)

Scientists use the following analogy to visualize how the expansion of our infinite universe might occur.

Suppose that we are baking an infinitely large loaf of raisin bread (infinite in extent as defined above, no edges.) We have a very large batter (the space of our universe) populated with an approximately uniform scattering of raisins (galaxies). We are in one of these raisins (the Milky Way galaxy). All that we see in any direction is other raisins, and the distribution of raisins as far as we can see in any direction is about the same. Our sight limitation defines a sphere that we will call our "observed universe." We believe that someone sitting on another raisin would have pretty much the same set of observations and their own "observed universe."

The batter (space) is expanding, causing the raisins to (in general) move away from each other. Those raisins already farthest apart move away from each other at the greatest speed, and those galaxies farthest from us move away from us at the greatest speed that we see. (However, raisins may also move slowly *through* the batter and in different directions. Locally, where the raisins near to us are being carried away by expansion very slowly, those with greater speed of movement through the batter toward us can actually be approaching.)

We postulate the following, based on the information supplied in our orientation earlier and Hubble's findings (with reference to our raisin-bread analogy):

1. The space (actually, space-time) of the universe is uniformly expanding at an accelerating rate, and its contents are (except locally) spreading apart because of that.
2. The universe is infinite in extent—no edges.
3. There is no center or preferred point of observation.
4. We can only know for sure what is within our sight, that is, an observed part of "*the* universe," what we will henceforth call "*our* universe."

Models of Our Universe

Until 1964, there seemed to be two strongly competing models of our universe that might explain Hubble's observations. *There have been many models over time. We describe these two and keep one.* Both were structured in conformance with general relativity and the key observations and principles listed and presented above.

> As defined by Peter Coles (writing as professor of astrophysics at the University of Nottingham), theories are self-contained with no free parameters, but a model is not complete in this sense.[14] To the extent that the model predicts what we actually observe, it may be considered successful.

The "Steady State" Model

This model was steadfastly defended by astronomer Fred Hoyle, among others.[15] It went beyond the cosmological principle to a generalization called the "perfect cosmological principle." Not only would the universe be without center anywhere in space, it would also be without a center with respect to time. The universe would be constantly expanding but always at the same rate. This process is called "continuous creation."[16] It would require the introduction of new matter at a steady rate to counteract the dilution caused by steady expansion.[17]

The "Big Bang" Model

This model differs from the steady state model in that the nature and rate of expansion of our universe substantially changes with time. However, as Coles points out: "Owing to the uncertain initial stages of the Big Bang, it is difficult to make cast-iron predictions and it is consequently not easy to test."[18] Nevertheless, for reasons that will be pointed out later, this is the model generally accepted today, and the one that we will use to guide us on our journey. So I describe it in much more detail below. (I understand that the label "big bang" was coined derisively by Hoyle in the heat of the argument over which of these models was correct.)

C. THE BIG BANG MODEL

Expansion and Cooling

We don't know what space-time was like at the beginning, we just speculate about what happened a short time after the beginning. The simplest models of the universe are of the big bang as a singularity (a point at which the mathematics predicts unrealistic infinities).

According to the big bang model, space-time at its beginning was imbued with an exceedingly hot, dense fixed amount of energy and fundamental particulate matter and antimatter (to be described later).[19] This universe has expanded and cooled over a stretch of 13.7 billion years from a temperature of trillions of degrees or more at the beginning, at the big bang. The cooling has allowed an "evolution" through a complex of processes to produce additional particles and the constituents of the universe that we see around us today.

As defined by Toback: "the evolution of the universe—is essentially the story of two important facts: (1) as the cosmos expands, the energy of the particles in the universe drop; and (2) the way that the particles interact with each other critically depends on the energy of the particles."[20]

The cooling is manifested in expansion of the wavelengths of light (electromagnetic radiation) that occurs along with the expansion of the space of the universe that the light occupies. Longer-wavelength photons are lower in energy and show a cooler universe. They indicate the temperature of the matter with which they interact. (Remember the classical picture of electromagnetic waves and the definition of wavelength provided in Fig. A.1 of Appendix A.) The stretching of wavelength is illustrated schematically in Figure A.2, which shows high-energy, short-wavelength gamma rays at the bottom left (at essentially the same hot time of the big bang listed at the bottom right) stretched to the long-wavelength, low-energy cold microwaves near the top left (that we see today, as indicated at the top right).

Don't be fooled by the seemingly small amount of stretching shown at the far left. This is just schematic. Moving from gamma rays (with a wavelength of 10^{-12} m) to

> microwaves (with a wavelength of 10^{-2} m) means the wavelength has expanded *ten billion times, which means our universe, correspondingly, has expanded ten billion times as well!*)

The results of this stretching can be reported in terms of temperatures, as explained in the indented paragraphs below.

> Electromagnetic radiation in the early universe would be constantly interacting with the particulate matter in its presence, being absorbed and reemitted in a constant exchange of energy. The spectrum of radiation in such an energy exchange, such a state of "thermal equilibrium," has a particular shape indicative of the temperature of the matter with which the radiation is interacting. (It was the derivation of a theory to explain this spectrum of radiation that led Planck to postulate that light would be emitted in "quanta," as was described in Chapter 2.)
>
> So the temperature of matter in the universe is indicated by the spectrum of radiation at and around a particular wavelength. The wavelengths of electromagnetic waves in thermal equilibrium with particles of matter can thus be used as measures of the temperature of the matter involved.
>
> It is customary in physics to use a Kelvin scale, which measures from the theoretical absolute zero of temperature, the lowest temperature that could ever be achieved. On this scale, the melting and boiling points of water occur, respectively, at 273 Kelvin degrees and 373 Kelvin degrees above absolute zero. Since water freezes at zero degrees Celsius, this means that absolute zero is at –273 degrees Celsius, or at –459 degrees on the Fahrenheit scale more commonly used in the United States. Where not otherwise specified, we'll be using a Celsius scale.

So, high-energy, short wavelength gamma-ray photons released just after the big bang would be characteristic of the trillions of degrees of the hot, dense universe at that time. And the low-energy, long wavelength, cosmic background microwave photons that we see today are indicative of the cool temperatures of deep space now, temperatures in the neighborhood of three degrees Kelvin—that is, three degrees above absolute zero, –270 degrees Celsius, or –453 degrees Fahrenheit. Cold! (We'll see later how these microwaves were found and examined.)

Maps of Expansion and Events

Phases of Expansion

The big bang expansion of our observed universe is depicted qualitatively in the sketch of Figure 9.1. In this figure, the largest observed size of the universe is projected forward and backward in time from what we observe now of its size as it was 0.5 billion years after the big bang[21] (the point on the curve to the lower left of the diagram). Half of the 0.16 million-billion-billion-mile diameter[22] shown at that point, is what we observe now to be the radius of the universe at that time. That radius is the distance *then* to the oldest stars and the farthest galaxies that we see now. We also know how fast the universe was expanding then, from a redshifting of the light from the stars in those distant galaxies.

We see these galaxies only now, because it has taken 13.2 billion years for light from those galaxies to reach us. That is why the rest of the curve, which indicates the growing distance of those same galaxies, is only a projection. The light leaving those same galaxies later than 0.5 billion years after the big bang hasn't reached us (and, as you will come to see, will never do so). And so the point at the upper right is only a projection of how that distance and size of the universe observed then will have grown until today.[23]

The overall expansion since the big bang is divided into *four phases*:

1. *Near uniformity*,
2. *(Cosmic) Inflation*,
3. *Slow Deceleration*, and
4. *Acceleration*.

(Note: The labels for the second and fourth phases are in common use. I have coined the terms to describe the first and third phases. All but the near uniformity phase are visible in Fig. 9.1.)

The *acceleration* phase, during the last six billion years, is shown in the increasing slope of the curve at the upper right. We can tell that faraway galaxies are moving away from us at an increasing rate. Scientists suggest that this acceleration is driven by a mysterious "dark energy"[24] that will be discussed in Section III (B) later.

The *slow deceleration* phase shows the observed universe expanding at a nearly steady but slightly slowing rate.

The (cosmic) *inflation* phase is shown as the rapid rise of the curve at the lower left of Figure 9.1. Inflation is described next below and is better shown in Figure 9.2, which spreads out the very early times for better viewing.

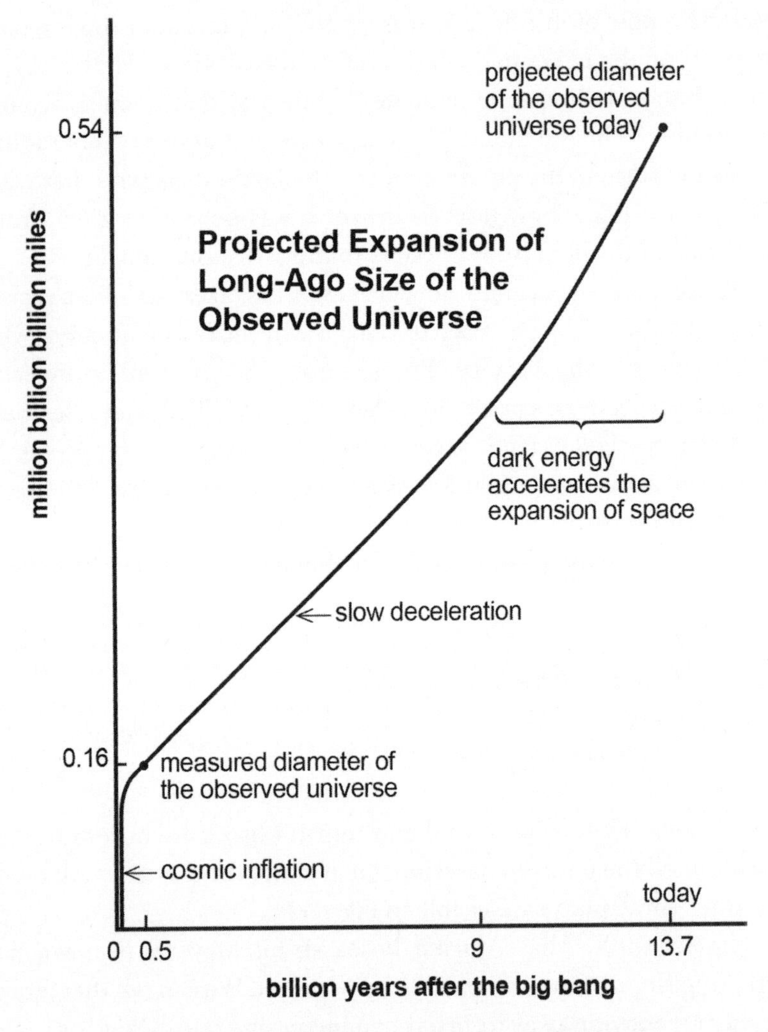

Fig. 9.1. Projected expansion of our observed universe.

The big bang model allows us to project backward from what we observe today of light from 0.5 billion years after the big bang (during the slow deceleration phase) to nearly the time of the big bang itself, suggesting that *inflation* would start a mind-boggling billion billion billion billionths of a second after the big bang and finish as early as 10,000 billion billion billion billionths of a second later.[25] (That's early and that's fast!) Inflation in that short time would bring the projection of our observed part of the universe nearly to the size shown millions of years later at the start of slow deceleration.

The *near uniformity* phase would appear before inflation. This phase will be described shortly, also in relation to Figure 9.2 (just ahead). But we need to do some homework first.

To easily handle small measurements of time, like 10,000 billion billion billion billionths of a second (that is, 0.000,000,000,000,000,000,000,000,000,000,010 seconds), and to make readable the labeling of the axes of Figure 9.2, we use a scientific notation. With this notation, for example, we write this very, very, very small time much more simply as 10^{-32} seconds. Since we will now be considering many more such small numbers in describing objects and events in time and space, and since we will also be considering huge distances and times, it seems appropriate to review again how we use scientific notation to simply describe both very small and very large numbers. And we'll include a review (once again) of a scientific shorthand for expressing physical relationships. Both methods will come in handy to make our reading easy in the pages ahead.

Remember first that the symbol *c* is used to refer to the speed of light, which we know from Einstein's formula $E = Mc^2$, for the energy equivalent of a mass of material. This formula, or equation, is just a scientific shorthand showing how *E* (energy) is related to *M* (mass) and *c* (the speed of light). Recall, $c = 299{,}793{,}000$ meters per second, where 299,793,000 is the number and "meters per second" is the set of units for speed, also written in shorthand as "m/s." (Remember that a meter, the standard international [SI] unit of length, is just three inches longer than a yard. You may be familiar with the terms *yardstick* or *meterstick* for the rulers used to measure length.) The superscript 2 after the *c* in the equation simply means that *c* is multiplied by itself: that is, $c \times c$.

To concisely express such large numbers as *c*, we'll be using scientific notation, in which we: (a) round off to the first few digits expressed in decimal form (2.998, for *c*) and then (b) multiply times the number ten raised to a superscript ("power") given by the total number of digits that would follow the decimal point from where we put it in the original number. (This is the same as the number 10 multiplied by itself that many times.) So, in this notation, $c = 2.998 \times 10^8$ m/s. Rounding off further, we would have the more easily remembered $c = 3 \times 10^8$ m/s. (Here 10^8 represents ten multiplied by itself eight times. We say that ten is "raised to the power eight." Each added multiple of ten is referred to as another "order of magnitude.")

We can express *c* in other units, for example *c* = 5.92 trillion miles per year, the number of miles traveled by light through space in a year's time. Scientists use the light-year, *c* × one year = 5.92 trillion miles, to describe very long distances. For example, ten light-years is ten times 5.92 trillion miles. It's just easier to use light-years, but miles is a unit that we can relate to, so I use it here.

We will also encounter some very small numbers: for example, as described earlier, the 0.000,000,000,000,000,000,000,000,000,000,010 seconds = 10,000 billion billion billion billionths of a second in which cosmic inflation occurs. (Note the thirty-one zeros before the digit 1, and then one more digit for the 1.) This number is just the number 1 divided thirty-two times by ten, or 1 divided by 10^{32}, more simply written as $1/10^{32}$. And this can be even more simply written as 1×10^{-32}, where the minus sign in the exponent means "divided by ten multiplied by itself thirty-two times." Or the number can be written still more simply as 10^{-32}, since 1 times any number is just the number itself. And, remember, 10^0 is just the number 1 (any nonzero value set to the zero power is 1. (Note that 0^0 is undefined.)

A Map that Expands Small Dimensions and Early Times

Figure 9.2 shows how matter coalesced after the big bang (lower curve, connecting the dots from left to right) against the backdrop of the projection of expansion in the size of the observed universe (upper curve).[26]

Special Scales

Note that Figure 9.2 is not your ordinary graph. Here we use what is called a *logarithmic scale* in powers of ten to compress the display of time and distance. This is done so as to show both the smallest and largest measurements of both time and distance reasonably on one page.

To get an idea of the compression, note that the top three divisions along the y-axis (vertical, showing distance) represent first a billion miles, then the hugely larger billion billion miles and finally a million billion billion miles. Similarly, the rightmost two divisions along the x-axis (horizontal, showing time) represent 10^7 (ten million) years and then the thousand times larger 10^{10} (ten billion) years.

The stretching out to show smaller and smaller sizes can be seen in the ten-billion-times (10^{10}) reduction in meters for each mark along the vertical axis, starting with the 10^{10} meters mark just below the division for a billion miles, and proceeding downward beyond 10^{-30} meters to include the Planck length of 1.6×10^{-35} meters, represented by the plus sign at the bottom left of the figure. This is the theoretical minimum increment in the composition of space.

Similarly, the stretching out of the display of shorter and shorter times can be seen in the ten-billion-times (10^{10}) reduction of seconds for each mark along the horizontal axis, starting with the 10^{10} seconds mark located just below the division for 10^7 years, and proceeding leftward to include the Planck time of 5.4×10^{-44} seconds, indicated by the same plus sign at the bottom left of the figure. This is the theoretical minimum increment in the composition of time. These tiny, tiny Planck increments in space and time have importance in the study of quantum gravity, string theory, and the earliest, densest and hottest moments after the big bang, as will be discussed later on.

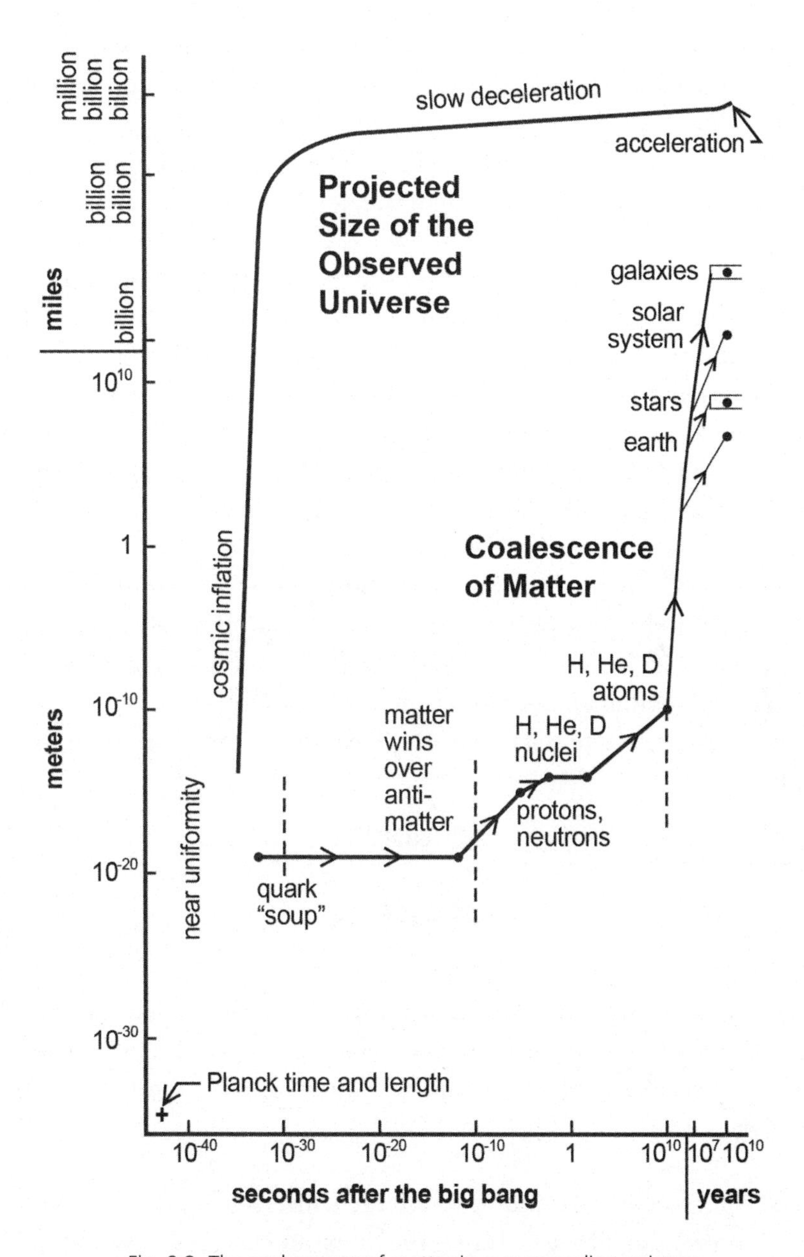

Fig. 9.2. The coalescence of matter in our expanding universe.

Note that the sharp cosmic inflation rise shown at the left of the figure curves over into a greatly decreased and slowly decreasing rate of slow deceleration, which (because of the extreme compression of large distance and expansion of early time in the scaling of this figure) appears as almost flat and unchanging.

To the left of the cosmic inflation shown in Figure 9.2 is the region of the first phase, *near uniformity*. Here the hot, first, crowded constituents of the universe would interact all with each other and achieve a near uniformity of distribution and a near thermal equilibrium (near uniformity in temperature). In the second phase, *inflation* would maintain this relative uniformity as it rapidly and substantially expanded the universe and the spacing of its constituents. Were this near uniformity of the first phase not there before inflation, it could never be established afterward, because a redistribution later would require that both particles and light move through space at speeds faster than the speed of light—which is impossible—to keep up with the expansion of space.

These two phases, near uniformity and (cosmic) inflation, together would seem to be necessary to explain the overall broad-brush uniformity in the distribution of mass and energy that we see throughout the universe today. (Remember the raisin-bread model.)

We now conclude Section I. We have our "maps." It's time to go sightseeing!

Section II

Sightseeing

(A Quick Tour through the Evolving Universe, from the Big Bang and the Very Small to the Present and the Very Large, with Stops afterward to Explore Particular Sights, including Those Topics in the Title to This Chapter)

A. Sightseeing Bus Tour through the Universe
B. Formation of the Galaxies, the Stars, and Our Solar System
C. Black Holes, Their Evaporation, and Gravity Waves

We get on the bus for a quick tour of the universe.

A. SIGHTSEEING BUS TOUR THROUGH THE UNIVERSE (OUR UNIVERSE FROM THE BIG BANG UNTIL NOW)

This is analogous to taking a quick sightseeing bus tour of a city. As often is the case, the bus driver is also the tour guide. He provides a bit of background information as we start out, so that we better understand the sights in the context of the overall layout of the city.

Stages in the Coalescence of Particles

Note the three dashed vertical lines in the bottom third of Figure 9.2. These indicate not only certain events (to be described later) but also three regions of our trip. The fourth region, to the right of the rightmost dashed line, includes the elements, the earth, the stars, the galaxies—stuff pretty much familiar to us—and the atom as described on a deeper level in the first seven chapters of this book and Part Four to follow. Between that line and the middle dashed line is the realm of nuclear and particle physics, revealed to us through experiments first using "atom smashers," ever larger and more sophisticated particle accelerators that have been built to break apart first the atom and then the nucleus of the atom into their constituent parts.

To the left of the middle dashed line in Figure 9.2 is a realm in which particle accelerators have created collision energies simulating the hotter and hotter conditions of the universe closer and closer to the early moments after the big bang. These experiments, guided by theory and then providing the bases for new theory, have revealed the fundamental particles that we now know. The process has led to the highly successful theoretical Standard Model that includes and describes these particles. (We discuss these particles, these accelerators, and the Standard Model in a later section of this chapter.) To the left of the leftmost dashed line, and particularly beyond the point showing the quark "soup," we enter a region where what happens gets to be speculative.

Now, the bus driver begins to describe some of the sights as we drive past them.

We trace the coalescence of matter as the universe expands and cools from the near the beginning that is the big bang. We start at the left end of the lower curve with the quark "soup."

Quark "Soup"

This "soup," containing nature's building blocks, is expected to have formed nearly uniformly throughout space at about 10^{-35} seconds after the big bang.[27] It would consist of a very, very high temperature mix of energy, electromagnetic radiation, and fundamental particles of matter and antimatter—the stuff that in stages eventually make atoms and the substances, stars, and galaxies that we see around us. (Most of these particles don't make up things we see today. Included were a lot of second- and third-generation quarks and leptons and probably a lot of neutrinos that are still streaming through space, and perhaps dark matter. We'll stop to take a look at these fundamental particles later on.) According to Turner (to be introduced shortly), it appears that this quark soup was also "the birthplace of dark matter,"[28] to be discussed further ahead.

Antimatter

For every particle of matter, there is an antimatter equivalent. And matter and antimatter have a dangerously explosive relationship. When matter combines with its antimatter equivalent, for example, a (negatively charged) electron with a (positively charged) positron, they annihilate each other and leave behind a huge (for those two tiny particles) amount of energy. (Remember, $E = Mc^2$, and c is a large number, 3×10^8 m/sec.)

Nobel laureate Richard Feynman, in his Feynman Lectures on Physics presented at Cal Tech over 1961 and 1962, jokingly made an interesting observation. Feynman pointed out that there should also be antiprotons and antineutrons, which together with the positron could, in principle at least, produce anti-atoms. While there might be a certain right-handedness to the direction of spin for the electron in the atom, it would necessarily be opposite, left-handed, in the anti-atom. If one of the anti-atoms should encounter an atom, the two would annihilate each other with the release of an enormous amount of energy (maybe ten thousand times the energy released from the electron-positron annihilation, because of the much greater masses of the atoms being converted to energy).

Feynman went on to reason that if there were anti-atoms, there should also be antisubstances, and antipeople. He wondered what would happen if aliens, perhaps Martians, had been constructed of antimatter: "What would happen when, after much conversation back and forth, we have each taught the other to make space ships [*sic*] and we meet halfway in empty space? We have instructed each other in our traditions, and so forth, and the two of us come rushing out to shake hands. Well, if he puts out his left hand, watch out!"[29]

Current theory states that matter and antimatter condensed out of the energy of the big bang in nearly equal proportions, but that matter had a minuscule edge; for normal matter, this would mean one extra quark for every billion antiquarks. (We'll learn about quarks and antiquarks toward the end of this chapter.) So when, at about 10^{-11} seconds after the big bang, the universe had cooled sufficiently that matter and antimatter could interact, most of the matter and all of the antimatter annihilated each other into energy. That minuscule greater amount of matter that was left over formed the normal matter of the universe that we see today.

Protons, Neutrons, and Atoms

(Here we follow the narrative of Michael S. Turner, theoretical cosmologist and distinguished professor at the University of Chicago, who coined the term "dark energy.") As Turner notes in his article titled "Origin of the Universe,"[30]at about 10^{-6} seconds after the big bang, after the universe had cooled to about a trillion degrees, "up" and "down" quarks formed protons and neutrons. These, during the span of 0.01 to 300 seconds, together formed the nuclei of the lighter elements, including hydrogen (H); helium (He); and the heavy isotope of hydrogen, deuterium (D), consisting of one proton bound to one neutron. Around 350,000 years after the big bang, these nuclei combined with electrons still present from the time of the "soup" to make charge-neutral atoms of the lighter elements. ("Charge-neutral" as distinct from the charged ions formed from atoms that have one or more electrons missing, stripped away, or added.) The combining of nuclei and electrons (a process called *recombination*) removed the free electrons that had been scattering light radiation, an event marked by the third vertical dashed line from the left, allowing the light throughout the universe to travel freely without scattering from the electrons. Wavelengths then continued to expand with the expansion of space, to produce the cosmic microwave background radiation that we observe throughout the universe today. (I'll describe this radiation and its significance toward the end of this chapter.)

Under the mutual attraction of gravity, the clumping of both dark matter and the atoms of these light elements then led to the formation of the galaxies and the stars within them. There is much to this process, including the formation of the planets, and we examine it more closely just ahead.

Energy and Matter

Realize that we have so far been describing mainly "normal matter," which constitutes only 4.5 percent of the total of all matter and energy in the universe. Of this, 0.5 percent is found in the stars and the planets, and 4 percent in a more widely distributed gas of molecules.[31] While normal matter consists overwhelmingly of hydrogen, with some helium, it also includes much smaller amounts of all of the rest of the elements (e.g., carbon, oxygen, nitrogen, iron, gold, etc.) and all of their constituent particles, many of them particles of the Standard Model, which will be described near the end of this chapter. Another 23 percent of the universe is dark matter,[32] also described toward the end of this chapter. It forms an unseen "halo" around the core of each of the galaxies. That halo is believed to hold the galaxies together, and by its distribution to be responsible for the peculiarly high speeds of the outer stars in their rotation around the cores of spiral galaxies. The remaining roughly 72 percent of the total of matter and energy is the mysterious "dark energy," believed to be responsible for an accelerated expansion of the universe displayed toward the top right of Figure 9.1 (as briefly discussed earlier and more fully discussed later in Section III (B)).

> The bus driver lets us off now in the first of a series of stops at points of special interest, some of them the topics in the title of this chapter. Local guides take over at each of these places on the tour.

B. FORMATION OF THE GALAXIES, THE STARS, AND OUR SOLAR SYSTEM

The Seeds of Galaxy and Star Formation

Professor David Toback[33] (who wrote our foreword) uses a playful analogy to explain the clumping of atoms toward the formation of the galaxies and the stars that we see today. In his description, gravity is recognized as a distortion of space-time according to Einstein's theory of general relativity, a "dent" in the universe of four-dimensional space-time caused by the presence of a mass of normal matter of any size, even the mass of an atom.[34] (Again, for our purposes here, we can just think of a dent in space.)

> We visualize the initial collecting of large amounts of matter to form a galaxy by thinking of little kids jumping on a trampoline. . . . Each child represents some mass (like an atom), and the trampoline represents space-time. As the kids jump, they create dents in the trampoline. If two kids collide, they fall into the trampoline and create an even bigger dent in space-time, making it more likely that other children will fall into that dent. As more kids fall, each dent gets bigger until everyone nearby falls into one of the giant dents.
>
> It is the same way with matter. Essentially, once you get a big dent in space-time, all the nearby matter starts falling into it. These big dents are where galaxies form.

If the universe were perfectly uniform to begin with, star and galaxy formation would have been an exceedingly long process, much longer than the 13.7-billion-year age of our universe.[35] But observations suggest that it took only half a billion years for the galaxies and stars to begin to materialize, which is a relatively short time. This would require a small local unevenness in the otherwise- even distribution of energy and particles in the *near uniformity* first phase of expansion just after the big bang. That unevenness, expanded, would later accelerate the clumping together of atoms and dark matter to make the galaxies and the stars within them. (More on dark matter farther ahead.)

The original unevenness is believed to have resulted from quantum fluctuations (to be explained later on in our discussion of black hole evaporation). But these fluctuations are calculated to occur only over tiny, tiny distances, also suggesting a compact, crowded universe in the initial *near uniformity* phase. Evidence for these quantum fluctuations is found today in a pervasive slight unevenness in the cosmic microwave background radiation (to be described later in this chapter).

Galaxy and Star Formation

The stars and the galaxies began to form at about the same time. Those expanded regions of space favored by the slightly greater initial energy and presence of mass, through gravitational attraction, would attract more and more matter as time went on, both regular matter (atoms) and dark matter (to be described later) to form clusters of galaxies. Within the clusters, regions favored even further by unevenness on a smaller but still considerable scale would attract atoms to form galaxies and the stars within them. (We'll talk more about the formation of stars and star evolution just a bit further ahead.)

Galaxies

Depending on circumstances, any of a number of types of galaxies may be formed. They all contain stars (which produce light), gas (just atoms) and mostly dark matter.[36] We consider here two types as examples: "spiral" galaxies and "elliptical" galaxies.

For spiral galaxies like our Milky Way (shown in Fig. 9.3, which is included in the photo insert), collisions of some of the normal matter in its movement toward the galaxy center would create a relatively dense clumping of a region of stars near the center. Other normal matter would begin to swirl, rotating more or less uniformly about the center while spreading out in a large "equatorial plane." The spiral arms of the galaxy are just a visual thing. They are places where there is more light, not more stars. The reason that there is more light is because the stars are younger there. The less-interactive dark matter would surround the center sphere in a very much larger unseen "halo."

The larger elliptical galaxies would more often form from the collisions of smaller galaxies. Stars in these galaxies and in the cores of the spiral galaxies travel in orbits having all manner of shapes and orientations. One such elliptical galaxy is shown in Figure 9.4, in the photo insert.

Stars

Stars were formed through the gathering of more and more atomic material, mainly isotopes of the hydrogen atom, with a heating up as the atoms became compressed more and more under the increasing gravitational attraction of a denser and denser overall mass. At first, the agitation caused by this heating would break loose the electrons from their nuclei, creating a "plasma" of positively charged ions and negatively charged electrons. Then, if the star were massive enough, on reaching a sufficient combination of temperature and pressure, the nuclei at the more compressed and hotter core region would begin to fuse together to produce the nuclei of heavier atoms, at first fusing hydrogen to produce helium. If the star were still more massive, this process would continue, producing additional successive concentric cores emerging at the center to (at various rates) produce the nuclei of heavier and heavier elements, such as lithium, beryllium, boron, carbon, nitrogen oxygen, and so on up to iron (with twenty-six protons and thirty neutrons in its nucleus).[37]

In each of these "fusion" reactions, the sum of the masses of nuclei that join together is greater than the mass of the nucleus that is produced, and the difference in mass, ΔM, is released as a tremendous amount of energy, as described by Einstein's formula for the equivalence of mass and energy: $E = \Delta Mc^2$. So the star heats up even more than would occur under the heating of compression caused by gravity.

The fusion process ends with iron because, as explained with the quantum mechanics of the nucleus (described briefly later on), it is difficult to add nucleons to iron. In fact, adding nucleons to make heavier elements than iron produces an increase in mass, absorbing rather than producing energy and shutting down the fusion process. Said another way, forming nuclei beyond iron requires the input of energy. Processes in stars of sufficient size provide the needed energy to create the rest of the elements of the periodic table, as we'll discuss further ahead.

As will be described in Chapter 20, efforts have been underway to create fusion reactors, little stars here on Earth, by combining the nuclei of hydrogen isotopes and using the energy released to power turbines and generate electrical power. Though born of the same understanding that produced the (fusion) H-bomb, fusion used this way promises safe nuclear fusion power, using hydrogen isotopes found in either freshwater or seawater as an inexpensive and virtually limitless supply of fuel.

Note that the nuclear reactors that we already have now in our "fission" power plants exploit this process in reverse, by splitting the nuclei of atoms of the heavier elements like uranium into smaller nuclei and particles (a fission process), and using the energy released to produce electrical power.

Pressure produced by the compression of the collapse itself and the added heating due to the release of energy in the fusion process counters the tendency toward collapse of the star under its own gravity. Depending on how this balance sorts itself out for stars of various total masses, and on the stage of consumption of the nuclei undergoing fusion, various stable or unstable types of stars and star remnants are produced. We can follow what happens for three different ranges of star masses, starting with the middle range of stars, stars that behave basically as our sun behaves.

Stars more than 8 percent of and less than eight times the mass of our sun all evolve in a somewhat similar manner.[38] They contract under the force of gravity until they heat up in a core region to about ten million degrees, at which point fusion begins. The added heating in the core, due to the fusion of hydrogen to form helium, creates a pressure that slows the collapse. Eventually an equilibrium is achieved at a fusion temperature and star size that depends on the star's mass. For our sun, that temperature is about fifteen million degrees, and the size is about what it is now, 872,000 miles in diameter, about 100 times the diameter of the earth. (The surface temperature of the sun is about 10,000 degrees.)

The star will stay in this equilibrium state until the hydrogen in the core is nearly consumed, at which point there is an initial collapse of the core under its own gravity as the expansive force of the heat of fusion reduces. The relatively quick collapse compresses the helium in the core to the order of a hundred million degrees, where the helium fuses to make beryllium and eventually the more stable carbon (even to make oxygen in stars having more than twice the sun's mass).

According to Gribbin, how this happens is "one of the most important discoveries in astrophysics (arguably, in the whole of science)."[39] In his book *13.8: The Quest to Find the True Age of the Universe and the Theory of Everything*, Gribbin writes:

> The problem is that beryllium-8 is unstable, and quickly splits apart to release two helium-4 nuclei (alpha particles). During the very brief time that a beryllium-8 nucleus (formed from a pair of colliding helium-4 nuclei) exists, it might be hit by another alpha particle; but it is so unstable that this ought to smash the beryllium-8 nucleus apart [not stick to it to make carbon-12]. And yet, if, hypothetically, beryllium-8 were stable, the calculations said that the production of carbon-12 would proceed so rapidly that the star would explode! Seemingly caught between the Devil and the deep blue sea, Hoyle found a way to strike a balance between the options of no carbon and too much carbon, provided that the carbon-12 nucleus has a property called a res-

> onance, with a very specific energy, 7.65 million electron volts [MeV], associated with it. . . .
>
> Hoyle convinced himself that, although there was no experimental evidence for such an excited state of carbon-12, it must exist.[40]

And he further convinced experimental physicist Willy Fowler and his team at Caltech to check it out, which they did.

> This was a sensational discovery whose importance cannot be overestimated. From the fact that carbon exists—indeed, from the fact that *we* exist—Hoyle had predicted what one of its key properties must be, and opened the way to a complete understanding of how the elements are manufactured inside stars.[41]

Fowler would in 1983 receive the Nobel Prize for Physics "*for his theoretical and experimental studies of the nuclear reaction of importance in the formation of the chemical elements in the universe.*"

The core's sudden further fusion and heating ignites fusion in the surrounding hydrogen shell, heating the rest of the surrounding hydrogen and propelling it outward in an expansion of the star to as much as one hundred times its initial size. The outer hydrogen cools in this expansion, taking on a longer-wavelength, lower-temperature, more red coloring in the light that it emits.[42] Hence the term *red giant* for this stage of the star's life, which lasts 0.5 billion to a billion years. When fusion of the helium in the core is complete and all that is left is an "ash" of carbon (and for larger stars also oxygen), the core experiences a second collapse and an intense heating that blows away all of the surrounding hydrogen. This leaves the core alone as a very dense, white hot, but slowly cooling *white dwarf*,[43] at perhaps a hundredth the diameter of the star to begin with, and with more than half of its original mass; such a star would be so dense that a piece of it the size of a sugar cube (½" × ½" × ½") might weigh several tons. Eventually, the white dwarf star will cool to the point that it will no longer emit any visible light, becoming in the end a *black dwarf*. But this cooling would take so long that no white dwarfs have ever reached this stage, and no black dwarf stars have been detected anywhere in our universe.

Our sun is 4.5 billion years old, and it is expected to convert to a red giant and then a white dwarf in another five billion years. (So we have a little while to plan for the future.) In general, the larger the mass of a star, the shorter its lifetime.

Stars less than 8 percent of the mass of our sun do not have enough mass to contract to a temperature that would start fusion. These stars are characterized as *brown dwarfs*, though strictly speaking they aren't stars at all, since their constituents don't undergo fusion.

Stars more than eight times the mass of our sun start out in a manner similar to their less massive brothers, but they have much shorter lifetimes and much more violent transformations at the end of their lives. Because of the much higher temperature created as one of these stars contracts under gravity, a succession of concentric cores form as the helium created from hydrogen is further fused to beryllium and then on to carbon, oxygen, and a succession of elements all the way to iron. Eventually all of the material in the core is converted to iron as the remaining "ash."[44] (The fusion process does not proceed to heavier elements than iron for the reasons described earlier.)

At this point, collapse occurs as it does for smaller stars, but the because of the greater mass of the larger stars, the collapse is not stopped by the mutual repulsion of electrons. The pressures in the cores are so great and the mix of particles so dense that electrons are pressed into combination with protons to produce neutrinos and an entire single core of neutrons. The intense gravity of the neutron core draws outer regions inward with such force that, combined with the energies of the neutrinos, a *supernova* explosion occurs that blows everything away except the neutron core. In the process, an entire spectrum of heavier elements including gold, platinum, and uranium are created. It is from an accumulation of these elements and everything else that is blown away that *the planets of solar systems*, such as our Earth, are created.[45] (Note that the dating of the oldest rocks found on our planet suggests that the earth has been around for about 4.6 billion years, about as long as our sun has been around.)

The light from such a supernova explosion can outshine for weeks or months the light emitted from small galaxies. The super-dense core that remains is called a *neutron star*.[46] Here the density boggles the mind. A sugar-cube piece of this star might weigh as much as the entire Earth. If this remaining neutron star has more than three solar masses, gravity quickly overcomes the mutual repulsion of neutrons and the star collapses into an entity of even higher density, a *black hole*. If the star is massive enough, it can develop directly from the supernova into a black hole, without the intermediate stage of the neutron star. Toback compares the supernova to a dandelion: "At the end of its life, a dandelion—like a giant star—turns from yellow to white, eventually goes 'poof!' and sends its seeds out to create the next generation."[47]

We make the next stop on our tour.

C. BLACK HOLES, THEIR EVAPORATION, AND GRAVITY WAVES

Black Holes and the Event Horizon

The "discovery" of black holes began with theory. Applying Einstein's general relativity theory to see what would happen to huge masses like stars under the force of their own gravity, the Russian astronomer Karl Schwarzschild in 1916 calculated that a sufficiently large mass would shrink down indefinitely in size to higher and higher densities, eventually reaching a singular point in space-time.[48] (These mathematically obtained singular points of infinite density and infinitesimal size, for both the big bang and black holes, cannot represent physical realities. The challenge for physicists is then to find what correctly describes what happens here, at these small sizes [where quantum mechanics should work] and these high densities [where relativity should work]. Unfortunately, so far, these theories seem to be incompatible. We discuss, later on, efforts to resolve the issue.) The prediction of black holes is more commonly attributed to Subrahmanyan Chandrasekhar in 1931,[49] who in

1983 was awarded the Nobel Prize in Physics *"for his theoretical studies of the physical processes of importance to the structure and evolution of the stars."*

The term *black hole*, coined by John Wheeler during a lecture in 1968,[50] derives from general relativity's prediction that no things, not even light, can escape from within a given region of such enormous gravity. Hawking describes the *black hole* as follows:

> Thus if light cannot escape, neither can anything else; everything is dragged back by the gravitational field. So one has a set of events, a region of space-time from which it is not possible to escape to reach a distant observer. This region is what we now call a black hole. Its boundary is called the event horizon and it coincides with the path of light rays that just fail to escape from the black hole.[51]

With regard to the last sentence, Toback explains the *event horizon* in a different way: "The *event horizon* for a black hole is the specific distance from the center where the velocity needed for escape is exactly the speed of light."[52]

According to general relativity, the trajectory of light is bent in the presence of a gravitational field. Inside the event horizon, gravity is so strong and the trajectory bent so severely that the light never leaves. And anything impinging from the outside and passing inside of this horizon is sucked into the hole, never to be seen again. Since nothing can escape from this region, particularly no light from the outside, the region inside of the event horizon appears to be black. The size of the event horizon and other characteristics of the hole's formation are all predictable from the mass, rotation, and net charge of the material that would form the hole.

Much is made in science fiction and in the movies of what might happen at the event horizon. Eventually an astronaut passing through the horizon would reach a place where his body would be stretched out and torn apart in a process that has been colorfully called "spaghettification." We can see that as follows: If, for example, he were to drop toward the center of the hole feet first, the much-greater gravity on his legs would pull on him far more strongly than the lesser gravity at his head. The difference in gravitational attraction would be so strong as to stretch him out and rip his legs from his body.

"Smaller Bangs" in Reverse?

In recent times, the black hole has taken on added importance as an object that can be examined to learn more of the extremely dense and extremely hot conditions that would have been experienced in the early days of our universe after the big bang. One might even think of the black-hole-like collapse of a massive object to a singular point in time and space as being a small model of the big bang in reverse:[53] the latter starting from a single, hot, energetic point in time and space and expanding and cooling to produce a coalescence of separate bits of matter as it grew in size.

Black-hole-like objects have been found, but only by inference from the way that they affect their surroundings. Being "black" (i.e., emitting and reflecting no light or matter), the objects simply cannot be seen directly. We classify them by the mass of energy and matter that they contain and the manner in which they affect their surroundings. I give here one example for each of two types of black holes.

"Stellar" black holes, the remnants of star collapse as described above, have been located from what would have been a pair of "binary" stars—except that one of the stars, now a remnant, a black hole, isn't seen. (The two objects rotate around like two dancers locked hand to hand and swinging around a point between them. But gravity, rather than locked hands, holds the two objects together in their rotation about the in-between point. The outward centrifugal force of rotation is countered by the inward attractive force of gravity.) These black holes usually have the order of several solar masses (where the mass of our sun is defined as one solar mass).

"Supermassive" black holes, with hundreds of thousands to billions of solar masses,[54] appear to provide the attractive force at the cores of spiral galaxies.

Perhaps the most visually striking evidence for a black hole comes from the light that is emitted from material as it is stolen from a nearby star and swallowed up or accreted into orbit around the hole, as shown in Figure 9.5, in the photo insert.

Gravity Waves

The recent detection of gravity waves, apparently from the merger of two black holes, may open the door to a new subfield of astronomy.[55] Such waves have been recognized as a theoretical possibility since Einstein's general relativity indicated the elasticity of space. It has been suggested that they may arise from cosmic inflation, supernovae, pairs of neutron stars, and pairs of black holes. A collision and merger of two such black holes might for a time place them in rotation about each other. This could create ripples in space-time at the frequency of their rotation, and these gravity waves would be expected to propagate outward like the ripples of water in a pond.

In September 2015, two very large and extremely sensitive pieces of apparatus located half a continent away from each other were able to detect these ripples, in what became the first direct detection of gravitational waves, and, apparently, the first detection of such binary black holes. Both devices concurrently recorded a nearly identical disturbance at just the frequency of rotation that would have been expected if gravitational waves existed.

The test not only opened up an entirely new mechanism for viewing the cosmos but also confirmed once again general relativity and revealed black hole masses at 29 and 36 times the mass of the sun, larger than the few solar masses typical of stellar black holes.[56]

While this is the first *direct* detection of gravity waves, there has been indirect evidence of their existence for some time. In 1974 the doctoral student Russell Hulse and his thesis advisor, Joseph Taylor, working at the Arecibo Observatory in Puerto Rico, discovered a rapidly rotating pulsar and a black companion star in a binary system having a loss of energy consistent with the emission of gravity waves. In 1993 they received the Nobel Prize in Physics *"for the discovery of a new kind of pulsar, a discovery that has opened up new possibilities for the study of gravitation."*

Black Hole Evaporation

Light *can* be emitted as a black hole evaporates. In 1973, thirty-one-year-old Stephen Hawking shocked an (at first) incredulous physics community with calculations that showed that quantum fluctuations in the space just outside the event horizon of a black hole would cause the black hole to evaporate. As he describes this in the bestselling book *A Brief History of Time*, such fluctuations occur all the time in what appears to be empty space.[57] Electric or gravitational fields, as described by quantum mechanics, can never really be truly zero, even as one might expect for nominally empty space. (If these fields were zero and unchanging, their rates of change would also be zero, and these certainties would violate the quantum physics embodied in Heisenberg's uncertainty principle.) The quantum nature of our world requires some uncertainty in the combination of the field and its rate of change.

So the fields fluctuate around a zero level. As you will see later on in the section on the forces of nature, fields are manifested in particles. So the presence of fluctuating fields will create short-lived pairs of "virtual particles," in particular, photons for the electromagnetic field. One particle of the pair would have a small positive energy while the other would have an equally small negative energy, so no net energy is created.[58] Normally these particles would quickly recombine to leave space pretty much as it was. But when such a pair is created near an event horizon, the particle of negative energy (having insufficient energy to escape, see Chapter 11) may be sucked inside, decreasing the total energy and therefore the total mass of the black hole, while its partner, having positive energy, escapes into space. (Remember that energy is equivalent to mass through Einstein's $E = Mc^2$.)

So, in theory, with billions upon billions of these negative-energy particles being captured and the positive-energy particles being released as "Hawking radiation" into the surrounding space, the black hole continually loses mass and shrinks in size at a rate (as Hawking calculated) that increases as the mass of the black hole gets smaller. Hawking suggested that in the end the last remnant of a black hole would disappear "in a final burst of emission equivalent to the explosion of millions of

H-bombs."[59] And if the virtual particles were photons, the escaping virtual photons would appear as if they were light escaping from the black hole. But, in fact, they would not have come from inside. Instead they would have come from just outside of the event horizon, thereby in no way violating the theory, which says that light cannot escape from *inside* the event horizon of a black hole.

Though other theorists have arrived at the same conclusions as Hawking, so the theory is pretty solid, black hole evaporation is very hard to find, and no one has actually observed or detected it yet. That is because black holes of substantial mass, say equivalent to that of our sun, radiate energy characteristic of objects at extremely low temperatures, on the order of 10^{-7} Kelvin degrees. The much "hotter" cosmic microwave background, characteristic of a temperature of 2.7 degrees Kelvin, pumps much more energy (equivalent to mass) into such a black hole than Hawking radiation removes. But a small black hole having the mass, for example, of an automobile would shine with luminosity two hundred times that of our sun, but it would evaporate in the order of a nanosecond (10^{-9} seconds), so you would have to be looking in the right place at the right moment. The search goes on: NASA's Fermi Gamma-Ray Space Telescope, launched in 2008, continues to search for these flashes, called gamma-ray bursts (GRBs).[60]

Even without experimental confirmation, the theory is so strong that many physicists take black hole evaporation as fact. That led to a serious problem that threatened the fundamentals of quantum theory.

Information Destroyed? Quantum Mechanics Threatened?

As discussed in Chapter 8, all physical things are information. All the contents of the black hole are information. That being the case, black holes appeared to threaten both information theory and quantum mechanics. Here's how.

As Leonard Susskind, professor of theoretical physics at Stanford, describes it in his bestselling *The Black Hole War*,[61] Hawking at a small meeting in 1981 dropped a "quiet bombshell"[62] on the physics community by observing that the virtual particles escaping from a black hole would not carry away any of the information stored within the hole. (After all, they come from outside of the hole.) As the black hole evaporates away and disappears, the information within the hole disappears also. It would be lost. This violates a key property and tenet of quantum mechanics (and information theory) called *unitarity*, which includes that quantum mechanics can move forward and backward in time (the essence of quantum logic) and that information, like energy, is always conserved and never lost.[63] (Relativity theory has no such unitarity tenet, and the basic relativistic description of the black hole would be unaffected.) This, more than anything that even Einstein and Schrödinger had put forward years before, might show quantum mechanics to be invalid or incomplete!

Susskind and colleagues eventually, over a period of sixteen years, came up with an answer to this challenge involving concepts well beyond the scope of this book: string theory (which I introduce only briefly later in this chapter), and the holographic principle.[64] It took another ten years for Hawking to accept that Susskind and company were right. (Hawking is apparently now a strong believer that the information in a black hole can be retained. In 2016 he published two articles in support of the idea, both in collaboration with colleagues Andrew Strominger of Harvard and Malcolm Perry of Cambridge University.[65])

> The last stop on our sightseeing tour is back at the planetarium. There, we examine the big bang model more closely.

Section III

Key Aspects of the Big Bang Model

A. Expansion of Space
B. Dark Matter and Dark Energy

A. EXPANSION OF SPACE

As you are by now well aware, the big bang model is built around the concept of expansion of space. We describe here the prediction and discovery of the cosmic microwave background, the key piece of evidence for that expansion. Peter Coles, then professor of astrophysics at the University of Nottingham and author of *Cosmology, A Short Introduction*, describes this as the "smoking gun that would decide in favor of the big bang model"[66] over the competing steady state model (that is described above in Section I (B)). In Coles's words: "The characteristic black-body spectrum of this radiation demonstrates beyond all reasonable doubt that it was produced in conditions of thermal equilibrium in the very early stages of the primordial fireball."[67] The steady state model in no way accommodates such a cooling (change of the universe with time), nor is it supported by many of the other pieces of evidence in Section III. This discovery of the microwave background is not only a critical test but also a fascinating story describing scientists and their instruments and some of the best of physics in an interplay between theory and experiment.

The Theory

The scientists George Gamow and Ralph Alpher (in the 1940s) were finding it difficult to theorize an accounting of the relatively high percentage of helium in the universe compared to what might have been produced by stellar processes alone.[68] Alpher and Robert Herman in 1948, then at the Applied Physics Laboratory at Johns Hopkins University, considered the idea that radiation at the early time of the formation of nuclei might have an influence on helium production, and that this radiation, expanded as a "cosmic microwave background" (CMB) might be detectable.

> A bit of scientist humor has been displayed in the contrived authorship of a paper by Alpher, [Hans] Bethe, and [George] Gamow, in a pun on the Greek letters "alpha," "beta" and "gamma." Gamow added Bethe's name (pronounced "beta") to the paper that he and his student were writing.

Actually, we now know that the light would come not from the formation of helium but from a later time and process called *recombination*.[69]

In the very hot early universe, free electrons would scatter electromagnetic radiation. But when the universe cooled from the trillions of degrees just after the big bang to the order of 3,000 degrees 380,000 years later, these electrons would be almost entirely captured in the formation of atoms. Electromagnetic waves, at that temperature light in the visible range of the spectrum, could then exist for reasonable lengths of time and travel freely. If their presence now were simply the result of such travel through space starting then, we would see short-wavelength visible radiation characteristic of the temperature at that time, perhaps redshifted a bit, but not radically. But if the space that these waves occupied were expanding greatly, then the wavelengths of this light would be expanded with it, on the order of several hundred thousand times, until what we find today—the increased wavelengths (reading upward in meters) in the first column of Figure A.2. (Note that time since the big bang [indicated in the last column] and the wavelengths [in the first column] are not spread according to any consistent scales.)

Alpher and Herman predicted that an expansion of space from that time to the present would stretch the wavelength of the radiation emitted at wavelengths typical of 10,000 degrees to a pervasive microwave background radiation characteristic of a temperature of about 5 degrees Kelvin. (Remember, the wavelength of light can indicate the temperature of its surroundings, as described in the section on expansion and cooling, Section I (C).)

How the Evidence Was Found

At the time that they proposed this expansion, the astronomical community was not particularly interested in examining theories of expansion. In fifteen years or so, that attitude changed, and apparatuses began to be constructed to test the theory. Meanwhile, Arno Penzias and Robert Wilson at the Bell Telephone Laboratories in Holmdel, New Jersey, stumbled upon the microwave background.[70] On May 20, 1964, they made measurements confirming a pervasive radiation corresponding to 4.2 degrees Kelvin (essentially what Alpher and Herman had predicted). Recent, more accurate measurements have pegged this number to be 2.75 degrees Kelvin.

The two men, Penzias and Wilson, had built a sensitive instrument intended for radio astronomy and were testing it. They didn't know that they were looking at CMB at all, thinking it was extraneous electronic noise in their communications system. They also thought it may have resulted from bird droppings in the externals of the apparatus. They only fully realized what they had measured when they took their findings to Robert Dicke, a noted physicist at Princeton who was actually building an apparatus to look for the CMB. When they published their findings, Penzias and Wilson did so alongside a parallel paper by Dicke on the interpretation of their results.

Fig. 9.6. Robert Wilson *(left)* and Arno Penzias *(right)* stand in front of their microwave horn antenna in Crawford Hill, New Jersey, in August 1965. (Image from AIP Emilio Segre Visual Archives, *Physics Today* Collection.)

Figure 9.6 shows Penzias and Wilson in front of the microwave antenna that they used to scoop-in and analyze the microwaves that they found. Their measurements were the same, regardless of the direction in which they pointed their apparatus. Those early measurements showed what appeared to be a uniformity in all directions of observed space. And because the spectrum of microwaves that were measured resemble what would be expected from radiation of cosmic origin, but with wavelengths stretched into the microwave range, their measurements support the big bang expansion of space as distinct from other expansion theories. In 1978, Penzias and Wilson would share the Nobel Prize in Physics "*for their discovery of cosmic microwave background radiation.*"

Quantum Fluctuations Accelerate Star and Galaxy Formation

Recent, More Precise Measurements

Recent, more precise measurements of the CMB radiation collected over a period of nine years (until 2012) show a slight unevenness in this background, which is indicative of a very small plus or minus 2×10^{-4} degrees Kelvin (about $4 \times 10^{-4} = 0.0004$ degrees Fahrenheit) unevenness in temperature. When this unevenness in temperature is converted into a variation in color, we see a "picture" of temperature variation throughout the observed universe like that shown as brighter to darker colors in the "sky map" of Figure 9.7(b), in the photo insert. This sky map has a construction similar to the map of the earth shown in Figure 9.7(a), except that the sky map is made from outward-looking observations and the map of the earth is from inward-looking observations.

Interpretation

This temperature map is evidence of the quantum nature of our world, of quantum fluctuations, even back to the time before cosmic inflation. The lighter and darker areas of this map are also consistent with the unevenness in the web of clusters of galaxies that we see today, in support of the suggestion that that acceleration of the formation of the stars and galaxies was ultimately caused by those quantum fluctuations. What is seen more directly in the CMB is an unevenness in the distribution of energy and particles at the time of recombination when the radiation observed today was released to be expanded in wavelength with the expansion of the universe. That relative unevenness, because of the magnitude and rapidity of inflation, would not have changed significantly from the unevenness that it grew from during the first near uniformity phase of expansion of our universe shown in Figure 9.2. And that unevenness at near uniformity is consistent with what could have been produced in that densely packed, small universe by quantum fluctuations. (Such fluctuations are discussed in Section II (A) in relation to black hole evaporation.)

The Galaxies

The web of galaxy clusters and the time of formation of the earliest stars and galaxies is consistent with the unevenness in the microwave background described above. And their movement away from us is consistent with the expansion of the universe as central to the big bang model.

B. DARK MATTER AND DARK ENERGY

As explained in Section I (A), the curvature and expansion of the universe depends on the density of the matter and energy in it compared to a critical density.[71] As Toback describes it, with the use of several methods of measurement, "our best estimate is that about 5% of the critical density is from known particles (with atoms making up most of the 5%), about 23% of the critical density is in dark matter, and 72% of the critical density is in dark energy"[72] that is believed to be responsible for an accelerated expansion of our universe.

Dark Matter

Dark matter may have formed in the quark "soup" in the early moments after the big bang.[73] As described below, its presence would explain the gravitational attraction needed to hold the galaxies together, and it would explain a number of other features of the galaxies. Dark matter isn't visible to us because it emits (and reflects) no radiation by which we can observe it.[74]

Dark matter, or something like it, seems to be necessary to explain gravitational effects on associated visible material. One of these effects is a strange, unexpectedly rapid motion of the outer stars and molecular clouds in rotation around the cores of spiral galaxies.[75] This motion indicates a substantial amount of unseen matter distributed widely throughout the region of rotation. Another is in "gravitational lensing,"[76] an effect predicted by Orest Khvolson in 1924,[77] and by Frantisek Link and, more famously, Einstein in 1936.[78] Light passing near a concentration of matter is bent in its trajectory in somewhat the manner that it would be bent in passing through an optical lens. Such a lensing around clusters of galaxies of light from faraway sources indicates the presence of far more matter than the normal matter seen in optical or radio telescopes.

From these and other effects, dark matter would appear to be distributed spherically around the cores of galaxies, with a density that lessens farther out from the core. This invisible spherical "halo" ascribed to dark matter is found to extend far beyond the visible halo of light from normal matter, suggesting different rates and mechanisms for the condensation of the two types of matter in the formation of the galaxies.[79]

Toback mentions that dark matter may be part of a new, more fundamental class of particles suggested by a possible supersymmetry, to be described a bit more in Section IV (C).[80]

Dark Energy and Accelerated Expansion

Dark energy is suggested as the reason for the apparent accelerated rate of expansion of our observed universe over the last six billion years.[81] We call dark energy "dark" simply because we can't see it and "energy" because it is only energy (as opposed to matter) that can be responsible for expansion.[82]

The acceleration has been calculated from the redshifting of the light from the explosions of distant Type Ia supernova. The measurements were carried out in the years preceding 1998 in what was called the Supernova Cosmology Project.[83] Saul Perlmutter, Brian P. Schmidt, and Adam G. Riess, all part of the project, in 2011 were awarded the Nobel Prize in Physics *"for the discovery of the accelerating expansion of the Universe through observations of distant supernovae."*

Section IV

Approaching the Big Bang

(Creating the Conditions of the Hot "Quark Soup"[84] just after the Big Bang, to Explore the Fundamental "Building Block" Particles of Nature and the Particles That Convey Nature's Forces)

A. Introduction
B. The World's Largest Single Machines—Colliders and Their Particle Detectors
C. Particles and Antiparticles of the Standard Model
D. Relativity and Quantum Mechanics Clash

In Section IV we witness the impressive interplay of experimental observation, analysis, and synthesis of mathematically expressed ideas that has given us a very satisfying, self-consistent view of the fundamental nature of our world, the Standard Model. Humans seek this kind of understanding and scientists feel a particular need for it. Here you get a flavor of the depth and breadth of accomplishment of modern physics and the questions that still remain.

A. INTRODUCTION

The possibility of breaking apart atoms and their constituents in search of the most elementary of particles has led for many years to our building various types of more and more powerful atom smashers and accelerators. But the prospect of creating these undiscovered particles out of pure energy in search of what may have existed at the very beginning of our universe after the big bang has in recent times led to the construction of accelerators of ever greater size, complexity, and power.

For example, Barnett et al., in their book *The Charm of Strange Quarks*, observe, "any pair of matching particle and antiparticle can be produced any time sufficient energy is available to provide the necessary mass-energy." (I'll describe antiparticles in Subsection C, just ahead.) As evidence they refer, for example, to a bubble-chamber photograph of the tracks of the creation of an electron/positron pair of particles from one high-energy photon. (You'll get a description of a more modern "drift chamber" in Subsection B.)

Remember Einstein's formula for the equivalence of mass and energy: $E = Mc^2$. Dividing both sides of this equation by c^2 still leaves this equation balanced and gives us $E/c^2 = M$, which is the same as $M = E/c^2$. Here M would be the combined mass of the electron and its antiparticle, the positron.

Colliders create beams of initial particles that are made to move in opposition at speeds approaching the speed of light.[85] Smashing them together creates new, more exotic particles to be examined. Among these are the fundamental "building block" particles of nature and the particles that convey nature's forces.

In Subsection C I will discuss the particles that have already been discovered, and other particles that may be expected. First, however, in this subsection (Subsection B), we examine how we find these particles, how we use colliders to create the hot, high-energy conditions that existed soon after the big bang. Here, we get a sense of the huge engineering effort that goes into making these massive and sophisticated instruments of particle physics. What we have learned with these machines is truly impressive, and, incidentally, a verification of quantum physics almost to the beginning of time.

Now our bus stops for a tour of an enormous and complex facility. We'll get to see hardware! But first the driver gives us an overview on these types of machines in general.

B. THE WORLD'S LARGEST SINGLE MACHINES—COLLIDERS AND THEIR PARTICLE DETECTORS

Colliders

Colliders have been built to maximize the amount of energy that goes into the creation of new particles. Because nature conserves momentum in any particular interaction, the net momentum of particles in the particle beams must be left in the products of collision. Any momentum left in products this way has its equivalent kinetic energy, and that energy does not contribute to the production of new particles. But if two particles can (ideally) be made to collide head-on with equal but opposite (in direction) momentum, because they are in opposition, there is no net momentum in their collision and no momentum or kinetic energy left afterward.[86] In this special case, all of the energy of the colliding particles goes into creating new particles. Few collisions are head-on, but enough collisions are close enough to head-on for a large fraction of the energy of the collision to be available to create new particles.

Negatively charged particles can be accelerated by attracting them toward regions of positive voltage, in somewhat the same way as was crudely done by Thomson in his discovery of the electron, as illustrated in Figure 2.5. (There, the electrons are accelerated from the cathode at negative voltage through the ring at high positive voltage before entering the collimating slits at A and B.) For particles of positive charge, like protons, the electric field would be reversed and they would be attracted toward negative voltages, and they would be accelerated many times through separate successive rings.

But billions of colliding particles are needed. Unlike other types of accelerators that can hit, for example, fixed solid targets where atoms are densely packed, beams-hitting-beams in colliders produce relatively few collisions. The particles are tiny. The chances of their hitting each other are small. (Imagine two fast-moving mists passing through each other vs. a single mist hitting a solid wall.) Typically, the particles in the beams are created and accelerated in "bunches."[87]

Barnett et al. describe the historical challenge of creating a sufficient number of particles in the quest to find the "W boson" of the "weak force." (This particle and this force are to be discussed in Section IV (C), to follow). Much of the accomplishment in high-energy physics is in building the accelerators to meet these kinds of needs. To create just ten W bosons would require a billion proton/antiproton collisions.

Carlo Rubbia, particle physicist and inventor, and technical physicist Simon van der Meer, with their team at the European Organization for Nuclear Research (CERN) near Geneva, devised and engineered the Super Proton Synchrotron (SPS) not only to accelerate protons and antiprotons to the energies required to create the bosons but also to provide an ample quantity of W bosons for detection and study. In 1984, they were awarded the Nobel Prize in Physics *"for their decisive contributions to the large project, which led to the discovery of the field particles W and Z, communicators of the weak interaction."*

(The SPS is now used as an injector, a preliminary accelerator to get particles partway up to speed, for the much larger and much higher energy Large Hadron Collider, which will be described shortly).

In colliders, the trajectory of particles and particle beams is controlled using a basic property of nature: charged particles are deflected sideways when traveling through a magnetic field. If the area of the field is large enough, the trajectory of the particles can be continuously bent in an arc that closes on itself. So charged particles can be made to move in a circle. Particles of opposite charge (let's say + and –) can be made to move in opposite directions on essentially the same circular path using the same magnets. As particles increase their velocity, particularly into the high-speed relativistic range approaching the speed of light, the magnetic field strength has to be increased to keep them on their circular path. The magnets that bend the trajectory of the beam by providing a vertical magnetic field are called "dipole" magnets. The particles in the beams repel each other electrically, which is why keeping them focused requires an inward magnetic force toward the beam axis. Magnets producing a field configuration that provides that force and keeps the beam narrow and focused are called "quadrupole" magnets. These are placed in a regular fashion in line with the dipole magnets periodically around the accelerator ring.

Let's suppose a group of particles all of the same type and charge are created by some means, injected in localized groups, and accelerated to higher and higher speeds in the presence of the increasing magnetic field, to circle in one direction. Colliders accelerating particles of opposite charge circulate two such beams (each having a fairly localized group of particles of one charge) in opposite directions. Both beams are accelerated with identically increasing speeds, guided on their circular paths by an increasing field strength. Finally, the two beams are brought together so that their high-speed particles collide in opposition at the location of an appropriate detector that records what happens. (The Large Hadron Collider, to be described shortly, creates two beams of like charge accelerated in opposite directions around essentially the same circular track with the trajectory of each beam controlled by its own set of magnets.)

How Colliders Are Constructed

Now here is where the engineering comes in. In practice, the trajectories of the particle beams have come to be circular and miles in diameter. (Strictly speaking, they are not circles but a series of arcs joined by very short linear accelerator sections.) The beams are enclosed in evacuated tubes so that no other particles will interfere with their passage. The tubes, which are permeable to the magnetic field, are enclosed in a chain of very long, powerful electromagnets linked end to end around the entire circle. As these beams of particles are accelerated to higher and higher speeds, these magnets produce an increasing vertical magnetic field to guide the beam essentially all of the way around the circle. For control, to get the peak strength of field required, and to reduce the power needed to drive them to reasonable field levels, all of the magnets are wound using superconducting wire. (You will learn more about superconductors in Chapter 19 and in Chapter 24.) To operate in the superconducting state (that is, with zero electrical resistance) to these field levels, the magnets are maintained at a very low temperature, just a couple of degrees above the absolute zero of temperature, by having them enclosed in what are essentially Thermos bottles filled or otherwise cooled inside with liquid helium.

A cross section through one of the dipole-magnet sections of the Large Hadron Collider (LHC) is shown in Figure 9.8, in the photo insert. (Note that "Hadron" derives from the Greek word *hadrós*, meaning "stout" or "thick." It describes a family of composite particles composed of quarks held together by the "strong" force, as will be discussed farther ahead.)

In Figure 9.8: (1) The two (colored white) "pupils" of what appear to be "eyes" in the figure are the evacuated tubes through which the particle beams travel. (2) The pair of (colored brown) "irises" are the cross sections of the two sides of the dipole magnet's superconducting windings. (To accelerate the protons in the two beams in opposite directions, the magnetic field from one iris is up while the other is down.) The superconductors are actually small, rectangular cables of superconducting wires, each wound inside of one of the many (colored brown) sectors of the "irises," traveling, say, in a direction that would be seen here as into the paper, the entire length of the forty-five-foot-long magnet in one sector to the right side of one of the tubes and returning in the equivalent sector on the left side, and around that loop again and again in that manner (crossovers at the ends of the magnets are not shown). (3) To keep magnetic forces from spreading the windings apart, they are enclosed in the (colored green) nonmagnetic, electrically insulating collars and "yoked" by (colored yellow) iron plates. (4) That assembly is enclosed in the inner lining of the Thermos-like inner (colored dark blue) structure. (5) The entire assembly is cooled by superfluid liquid helium at 1.9 degrees Kelvin, that is, just 1.9 degrees Celsius [or 3.5 degrees Fahrenheit] above the absolute zero of temperature, which, as noted earlier, is at –273 degrees Celsius, or at –459 degrees Fahrenheit. (Superfluid helium conducts heat without resistance in much the same way that superconductors conduct electrical current without resistance.) (6) The rest of the inner structure consists of a (colored tan) superinsulation (to reflect heat radiating in from the outside) between the inner and outer (colored light blue) cylinders of the Thermos bottle. (The design of cold components is in the realm of cryogenic engineering—liquid helium, liquid hydrogen, and liquid nitrogen, for example, being cryogens.)

Because the beams of charged particles are accelerated forward to increased speed, and accelerated sideways to stay on their circular trajectories, they emit radiation—lots of it. (Such acceleration is illustrated in Fig. 2.6 and discussed in nearby paragraphs. Remember, this acceleration and radiation is why the Bohr model of atom was flawed, as discussed in Chapter 2.) So, as a safety measure to protect both life and equipment during operation of the collider, the entire ring of tubes, magnets, and cryogenic enclosures is assembled and housed in an underground circular tunnel, appropriately enclosed for protection against the elements. Figure 9.9 shows the assembly of Figure 9.8 inside the tunnel. Figure 9.10 sketches the layout of the LHC ring, its experiment stations, and its pre-accelerators. (All three figures can be found in the photo insert.)

A Comparison of the Latest and Future Colliders

Five generations of the most powerful recent, presently operating, and planned[88] colliders are listed below.[89] These have already been introduced or will be introduced and/or described at various points in the remaining sections of this chapter. (Not shown in the table is the Superconducting Super Collider, which would have been much larger and more powerful than LHC. Construction was begun in Texas, but the project was canceled during the Reagan administration because it came to be viewed as too expensive.)

Collider	Where	Operation	Accelerated	Max. TeV	D, Miles
Tevatron	USA	1987–2011	proton & anti	1	4.3
RHIC	USA	2000–	pp, AuAu, Cu	0.5	0.8
LHC	Switz.	2008–	pp, pPb, PbPb	6.5	5.3
?[90]	China	2028–	?	?	10
VLHC [91]	?	?	?	?	20

(The column labeled "Accelerated" displays colliding beams of: protons and antiprotons, pp [protons and protons], AuAu [gold ions and gold ions], CuCu [copper ions and copper ions], pPb [protons and lead ions], and PbPb [lead ions and lead ions].)

Many of the particles of the Standard Model of particle physics (to be described in the next section) were discovered using the Tevatron and other contemporary or earlier colliders and accelerators (not shown in the table above). The Tevatron, built at Fermilab near Chicago to smash protons and antiprotons together, was the most powerful operating accelerator in the period from 1987 until 2011, producing a maximum single-beam energy of one TeV, that is a million million electron volts. All in all, the construction of the main ring of the Tevatron in 1983 and then its upgrade in 1993 cost over $400 million.

Note that "Tevatron" derives from the collider's operation in the terra-eV (= million million electron volt) range of energy for each colliding beam of particles. Each particle in the beam can be made to achieve that energy before collision. As you will see with further discussion in Chapter 11, eV, the electron volt, is a commonly used measure of energy in atomic and particle physics. The energy of the violet photon emitted in transition of the electron from the $n = 5$ to the $n = 2$ state in the hydrogen atom (shown in Fig. 2.9) is approximately 3 eV. Five times that energy, 15 eV, will easily break an electron free of the atom. The Tevatron created particle energies more than 60 billion times larger, going beyond what is needed to split atoms, beyond the energy needed to break the nuclei of the atoms apart, beyond the energy needed to break their constituent protons and neutrons into their constituent quarks, and, finally, to the energy that is needed to create the (heaviest of all) top quark described in Figure 9.11 (in the photo insert), the last fundamental fermion to be discovered. (These particles are described in the following section of this chapter.)

The Large Hadron Collider

The latest particle collider to be built of the list above is the presently operating Large Hadron Collider. In 1998, construction was begun on the LHC at CERN near Geneva; at $4.4 billion, it is the most complex experimental facility ever built, and it is the largest single machine in the world. The LHC started operation in 2010, accelerating beams of protons in an underground tunnel that runs as deep as 574 feet below the surface, crossing under the French–Swiss border and then back through a circular trajectory that is over five miles in diameter. Opposing beams are collided to create and explore the behavior of particles at the energy densities experienced in the beginning fractions of a second of the existence of our universe, just after the big bang.

In all, 1,232 superconducting dipole magnets (each about forty-five feet long and weighing nearly forty tons) are positioned end to end to bend a beam of protons through its 5.3-mile-diameter accelerator ring. It circulates up to 2,800 bunches of protons with 118 billion protons in each bunch at up to 6.5 TeV per proton, for a total collision energy of up to 13 TeV per collision pair. At peak energy, the bunches of protons move at just 7 miles per hour slower than the speed of light[92] (which is 3×10^8 m/sec = 3×10^5 km/sec = roughly 2×10^5 miles per second, or 700 million miles per hour.

In 2012, using this facility, physicists found first evidence of a Higgs boson. (By that time, six quadrillion [6×10^{15}] LHC proton-proton collisions had been analyzed. Since that time, additional work has found the Higgs boson to behave in ways predicted by the Standard Model, including that it has a zero spin.

Although one justification for construction of the LHC was to find the Higgs boson, the long-sought predicted particle of the Higgs field that would give particles their masses and complete the Standard Model (all of this will be discussed in Section IV (C)), the maximum design energy of the LHC far exceeds what was thought to be necessary for the Higgs. The extra energy may have provided a cushion on that quest, but it is now also opening the door to new particles and new physics, taking us a big step closer to understanding the big bang.

Particle Detectors

Creating particles is one thing. Detecting them is another. There are many kinds of particle detectors, and these may be chosen according to the types of particles to be detected. Or they may be built specifically for a particular experiment.

In some of these, the high-energy particles produced in the accelerator strip away electrons from atoms in a gas as they pass through it, leaving a trail of charged ions and electrons. The gas is in an enclosure called a "chamber" in a magnetic field that bends the trajectory of any particle with charge. From their trajectories we can determine the particle's momentum and charge.

In the boxed text below, I quote part of a paragraph from the book by Barnett et al., to outline just one example of a commonly used detector.

> A drift chamber "contains many anode wires (wires at positive potential [voltage]), each surrounded by a grid of cathode wires (negative potential [voltage] relative to the anode wires). Each anode wire collects ions from the region between itself

and the nearest set of cathode wires."[93] "The 'drift time,' that is, the time taken for the ion to travel from its point of origination to the wire, is proportional to the minimum distance between the track and the anode wire."[94] And so, without going into further detail: from the measurement of this drift time for all of the ions created along the detected particle's path, we can determine where each ion started its drift, and then we can plot the particle's path.

Sometimes the sought particle isn't detectable directly, but it might have a "signature," for instance, decaying into other particles in a particular way that *can* be detected. A good description of the sophistication involved in particle detection is provided in the article: "The Higgs at Last."[95] (We'll learn about the Higgs in Section IV (C), just below.)

I won't go on. You get the idea—particle physics involves sophisticated experiments that require a lot of engineering and extensive construction (in addition to cadres of scientists, mathematical tools, and computer-automated tools to analyze massive amounts of data). But, what we have learned from these experiments is truly amazing, as you are about to discover.

C. PARTICLES OF THE STANDARD MODEL

In this Section IV (C), we make a very long story very, very short. Humankind has always been on a quest to determine what the substances of the universe are made of. The Greek word *atom* was used to describe the smallest piece of matter, something that literally could not be cut into smaller pieces. Starting in 1930, a "zoo" of subatomic particles was discovered. Today we know not only that atoms are composed of nuclei consisting of neutrons and protons surrounded by clouds of electrons, we know that the neutrons and protons are formed from different combinations of smaller particles called *quarks*. So far as we know today (although there is speculation to the contrary), the quarks and electrons are not composed of anything smaller, and so are considered *fundamental particles*.[96] There are many other fundamental particles, as will be described shortly. And, by 1970 most of the fundamental particles had been explained by theories that fit into what is called the Standard Model.

Now realize that all particles in our universe are of two general types.

Fermions[97] are particles having half-integral units of spin (i.e., plus or minus 1/2, 3/2, 5/2, and so on). In nature, no two fermions are found to occupy the same state. (Remember the electron with spin ½, and the Pauli exclusion principle.) This leads to their being describable by a "Fermi" type of statistics (named after the physicist Enrico Fermi).

Bosons[98] (named by Paul Dirac after the Bengali physicist Satyendra Bose[99]), are particles of integral spin (i.e., 0 or plus or minus 1, 2, 3, and so on). In nature, any number of bosons may be found to occupy the same state. This leads to them being describable by a "Bose" type of statistics.

Fundamental Particles and the Standard Model

Because of this universal division, the Standard Model includes two sets of "fundamental particles." "Fundamental fermion" particles comprise the first set. Every one of the twelve fundamental fermions has a spin of ½. All other particles, all matter, is made up of one or more of these fundamental fermions. "Fundamental boson" particles comprise the second set. Four of the five fundamental bosons convey the forces of the fields responsible for the interactions between fermions, otherwise described as "mediating interactions among fermions."[100] These four are called "gauge" bosons. The fifth fundamental boson is the Higgs boson. (More about the Higgs boson later on.)

The fundamental fermions and bosons have all been discovered in nature or created in the laboratory. They are listed in Figure 9.11, which is included in the photo insert.[101] Not included in Figure 9.11 is a possible fifth fundamental gauge boson, the "graviton," which would convey the force of gravity. But the graviton has not been found, nor has a theory been developed to formally project it. (More on the graviton toward the end of this chapter.)

At present, these fundamental fermions and bosons are all of the known fundamental particles. But it has been suggested that a "supersymmetry" exists that would have every fundamental fermion particle have a related fundamental boson particle. This would change our view of what is "fundamental," and it should involve the existence of new particles as yet undiscovered but possibly in the range of energies that can be provided by the LHC and its successors. (Perhaps some of these particles are what we call dark matter.) We'll just have to see. New theory may be required. Regardless, I proceed here with what most scientists accept for now as the fundamental particles and the theories of the Standard Model that describe them.

In the pages ahead I am going to present the fundamental fermion particles first and describe some of their properties. Then I'll provide a very short history of how fermions and bosons were discovered. Next, after that, we'll count the number of fundamental particles. And finally, we'll examine the collection of theories making up the Standard Model that explain the fundamental fermions and the bosons responsible for their interactions.

The two sets of fundamental particles (as far as we know) comprise all of the *fundamental* particles in the universe. Together they make up all of the particles of the *Standard Model*. The model both incorporates and describes them. There are other fermions and bosons, but they are made up of fundamental particles and are not shown or discussed here. The fundamental particles are listed in their fermion and boson groups in the table for the Standard Model shown in Figure 9.11, in the photo insert.

Note that the electrical charge for all of the fundamental particles is given in Figure 9.11 at the top right of the round icon labeling the particle. And note that the mass of every particle in the figure is given below the round icon for that particle in millions of electron volts, MeV. For now, to get a feel for how massive the various particles are, just realize that the electron mass is about half an MeV.

Fermions are shown in the first three columns of Figure 9.11. Each successive column contains a successively heavier "generation" of fermions. With their antiparticles (to be discussed shortly), fermions comprise all that has been found or created in our universe of what is called "normal matter."[102] (Dark matter was considered in Section III (C), above.)

Bosons are listed in the last column of the figure. The top four of these bosons convey the forces of nature, as will be described later in this section. The fifth boson, the Higgs boson, is different, as will also be described later.

Fermions

The fermions are listed in two major subgroups: the first subgroup consists of the two families of *quarks* (both shown in red): those in the first row with +2/3 electrical charge, and those in the second row with −1/3 electrical charge. The second subgroup contains the two families of *leptons* (both shown in green): those in the third row all have zero electrical charge, and those in the last row all have a –1 electrical charge, including the familiar *electron* at the bottom left corner of the table.

All normal matter found in nature today is made of quarks or leptons, or matter constructed from them. Protons are composed of two +2/3 charge "up" quarks and one –1/3 charge "down" quark, yielding the net +1 charge that we observe. Neutrons are composed of one +2/3 charge "up" quark and two –1/3 charge "down" quarks, yielding the net zero charge that we observe. Protons and neutrons form the nuclei of atoms, with a net positive charge given by the number of protons. Those nuclei, surrounded by an equal number of –1 charge *electrons* make up the charge-neutral atoms of the elements. (When electrons are stripped away or added, what results is ions of positive or negative charge that are part of the fascinating chemistry to be described in Part Four of this book.) The properties of each type of atom, of each element, are related in fascinating ways (to be described in Part Four of this book) to the specific number of electrons or equal number of protons that they contain.

Fermions, Fields, Forces, and Bosons in a Historical Context

The Strong Force

How could it be that a nucleus that consists of protons each having electrical charge (and neutrons having no net electrical charge) can stay tightly bound together? After all, like electrical charges repel. The protons should push each other apart. The answer is that the nucleus is bound by a *strong nuclear force, which is* stronger than the electrical repulsion. This force *was* proposed by the Japanese physicist Hideki Yukawa in 1935. According to Yukawa, this force would result from neutrons and protons exchanging subatomic particles of a kind that at that time had not yet been detected. In 1949, Yukawa became the first Japanese to be awarded a Nobel Prize in Physics, "*for his prediction of the existence of 'mesons' on the basis of theoretical work on nuclear forces.*" But this was an intermediate result. We now know that the mesons, protons, and neutrons themselves are composed of fundamental particles called quarks, which produce the strong force by exchanging gluons.

Quarks, Gluons, and the Photon

Particularly throughout the 1960s and culminating in 1970, scientists examined the "zoo" of subatomic particles by cataloging them according to their properties. In a manner similar to Mendeleev's finding the pattern of the periodic table that then led to the understanding of electrons and the quantum mechanics of the atom, Murray Gell-Mann explained a threefold symmetry in the properties of the particles by postulating (and then explaining them theoretically) an entire set of fundamental fermions and force-exchange-particle bosons.[103] To distinguish them he invented an imaginative new set of labels.

The fermions, he called quarks, borrowing the spelling from a line in *Finnegans Wake* by the Irish novelist James Joyce, "Three quarks for Muster Mark."[104] The bosons he called (appropriately, it would seem) *gluons*. Six quarks occupy the top, leftmost group (in red) of the particles of the Standard Model shown in Figure 9.11. The first three to be discovered were named (strangely) "up," "down" and "strange." The gluon is shown as the second boson in the last (blue) column, after the familiar boson associated with electromagnetic radiation, the *photon*—and, like the photon, it has no mass.

The Electro-Weak Force, the Neutrino and W and Z Bosons

In 1930, Wolfgang Pauli suggested that "beta decay," the observed transformation of a neutron into a proton and an electron, must be accompanied by the emission of an as-yet-unseen electrically neutral particle of ½ spin and having little or no mass.[105] Fermi in 1933 developed the theory of beta decay, describing the unseen particle as "the little neutral one," in his native Italian, the *neutrino*,[106] what we now know to be the first-generation "electron neutrino" lepton type of fermion shown in the table. It wouldn't be until 1956 that neutrinos were actually detected, inside of a nuclear reactor that produced scads of free neutrons that might decay.[107]

"QUARKS. NEUTRINOS. MESONS. ALL THOSE DAMN PARTICLES YOU CAN'T SEE. THAT'S WHAT DROVE ME TO DRINK. BUT NOW I CAN SEE THEM."

By 1970, theorists Sheldon Glashow, Abdus Salam, and Steven Weinberg (who would share the Nobel Prize for this), had combined the weak force with electromagnetism to produce the *electro-weak theory*, which included the *W and Z bosons* in conveyance of the force and predicted a fourth type of quark that they called *charm*.[108] The W and Z bosons are shown second and third in the last column of bosons in Figure 9.11.

The Higgs Boson

Three groups of theorists working in parallel in 1964 came up with theories that predicted an unusual field, constant almost everywhere, that would allow existing particles to acquire mass. This eventually became known as the "Higgs field," after one of the investigators, possibly because his theory described not only the field but also the associated particle. Their work was pretty much ignored initially. But as more and more particles began to be discovered during the 1970s, these theories came to be recognized and became one of the theories of the Standard Model.

Though the Higgs boson is not a carrier of force in the same sense as the other bosons, it is nevertheless listed with the gauge bosons of the Standard Model as the last particle in the last (blue) column of Figure 9.11.

Mass

Note again that the mass of every particle in Figure 9.11 is given below the round icon for that particle, as an energy in millions of electron volts, MeV. That is because of the equivalence of mass and energy—remember, $E = Mc^2$. (Again, just remember that the mass of the electron is about 0.5, and the numbers tell you how heavy the rest of the particles are compared to the electron.)

This conversion of energy to mass has already been discussed in the box dealing with colliders in Subsection B. It is interesting to note that the 0.511 MeV energy equivalent of the mass of an electron (shown under the icon for the electron at the lower left in Figure 9.11) is about thirty thousand times the roughly 15 eV energy needed to break an electron loose from the hydrogen atom. The latter is the sort of energy involved in chemical reactions. The former is the sort of energy involved in particle physics.

As noted before, any mass can be created out of pure energy. The top quark and the Higgs boson have masses the order of three hundred thousand times the mass of the electron. It takes a lot of energy to create these heavier particles. That is why the top quark was the last quark to be produced and discovered in the Tevatron, and why it has taken until the last couple of years and the LHC to produce the Higgs boson. Part of the energy delivered in collisions can come from the mass of the particles colliding. But most of the energy for the creation of new particles comes from the very high kinetic energy of the colliding particles, the energy of motion. With speeds for the colliding protons of the LHC approaching to within 7 miles per hour of the speed of light, about 700 million miles per hour, enormous energies are available.

"Color Charge"

Protons and neutrons are part of a family of particles called *baryons*, defined as "a hadron formed of three quarks"[109] reflecting the threefold symmetry mentioned above in the discussion of the strong force. And quarks have never been found alone. Murray Gell-Mann postulated and then showed that the quarks would have another property that he *called color charge*, or simply *color* (not to be confused with electrical charge or the color of light): either a "red charge," a "blue charge" or a "green charge" (named after the colors that combine to make the colors we see on our color television sets). The three color-charge quarks combine to produce the (exceedingly strong) strong force. Fred Bortz, physicist, science writer, and critic, describes this force: "Unlike gravity and electromagnetic forces, which decrease as particles move apart, the color force behaved more like a coil spring, pulling quarks together with increasing intensity the farther apart they became. This is the reason that bound quarks are inseparable."[110] Gell-Mann, by his labeling choices, made it so that the only way that quarks bind together is when they form a white composite. (Similarly, in a color TV, blue, green, and red make white.) The hadron may also contain additional quark/antiquark pairs,[111] which are also tightly bound together. One color charge and its anti–color charge together also produce white. Mesons are composed of one or multiple quark/antiquark pairs.

Antiparticles and the Number of Fundamental Particles

An *antiparticle* is predicted (and found) to have the same mass as its equivalent particle, but the opposite of all other particle characteristics; in particular, a positive electrical charge would become a negative electrical charge in the antiparticle.[112]

> For example, each of the fermions has a partner particle (also a fermion) that mirrors its properties (as described in part earlier in the discussion of the evaporation of black holes, and with Feynman's humorous description of the meeting of Martians in space). The first indication that there should be such a mirroring came through the theoretical work of Paul Dirac, when in 1930 he used quantum mechanics and relativity to derive, from just its mass and its unit of minus charge, all of the rest of the

properties of the isolated electron. His calculations also showed that there should be an antielectron with opposite (positive) charge.

This particle was found by Carl Anderson one year later in a cloud-chamber photograph of cosmic radiation. Dirac shared the Nobel Prize in Physics in 1933 "*for the discovery of new productive forms of atomic theory*," and Anderson was awarded the Prize in 1936 "*for his discovery of the positron*." Dirac's analysis, extended to the rest of the particles, would say that they all should have antiparticles, and these have also been observed as products created in particle accelerators.

Every particle of the Standard Model shown in Figure 9.11 has an antiparticle, so the total of fundamental particles is 34.

The Theories of the Standard Model

Whereas the big bang model has at its core general relativity, the fundamental particles of nature (now and as they existed in early times near to the big bang) are described through the quantum description of the fields of nature and their interactions. Each of the four fundamental fields interacts with matter through its associated gauge boson force carrier, those bosons listed first in the last column of Figure 9.11. The Higgs field is different in that its boson is not a purveyor of force and not a gauge boson.

Quantum mechanics applied to each of the fields is described below in order of its historical development. The first application to electromagnetism, is described in a little more detail to provide a sense of the way experimental and theoretical approaches came together.

(Remember: All of the fundamental particles and fields of the Standard Model have been found in nature or in particle accelerators. They behave as quantum entities. It's a quantum world. The theory just explains how that happens.)

Quantum Electrodynamics— Quantum Mechanics Applied to Electromagnetism

Recall that Bohr and Schrödinger developed their models of the hydrogen atom recognizing that a negatively charged electron would be attracted to the positive charge of the proton nucleus. Classical theory would say that the electron was attracted via the electric "field" that extended through space in all directions from the positive charge of the proton. Well, this view of how the forces would act, though useful for Schrödinger and still for most practical purposes, was changed with a further development of quantum mechanics, particularly, during a period of seven years starting in 1942, with an interactive effort including five physicists in the United States.[113] One scientist in Japan worked on this alone.

The work started in the United States when twenty-four-year-old Richard Feynman, in graduate school at Princeton—and, independently, another comparably young New Yorker, Julian Schwinger—tackled a longstanding problem. Previous attempts to use quantum mechanics predicted that the energy of a charged particle in its own electric field would become infinite (exhibit a singularity). The work was also spurred when Willis Lamb precisely measured the spectrum of atomic hydrogen and found a tiny shift in results that couldn't be explained by any previous theory (the "Lamb shift").

Schwinger would avoid the singularity using a mathematical process called *normalization*. Feynman realized that he could better describe the action of an electric field as resulting from *virtual photons*. In particular, the force between a proton (positive charge) and an electron (negative charge) would be viewed as a continual emission and absorption of a pair of virtual photons exchanged between the proton and the electron. (The photons are virtual because the sudden appearance of two real particles with mass [each equivalent to energy through $E = Mc^2$] would violate the conservation of energy. What allows the existence of the two virtual particles is that they are an "uncertainty" in energy, ΔE. As long as their lifetime, Δt, is short enough, they can exist as described by Heisenberg's uncertainty principle applied to energy: $\Delta E \times \Delta t < h$ [Planck's constant again]).

Both approaches and a third pursued independently by Japanese physicist Sin Itiro Tomonaga avoided the singularity and explained precisely the Lamb shift. In 1965, the three men received the Nobel Prize in Physics "*for their fundamental work in quantum electrodynamics, with deep-ploughing consequences for the physics of elementary particles.*" The resulting zero mass, spin = 1 photon, the gauge boson of the electromagnetic field, is listed as the first force particle in the last (blue) column of bosons listed in Figure 9.11.

In the course of his work, Feynman created simple diagrams to provide a good visual approach for conceptualizing particle interactions, and his summing-over-all-possible-paths method for calculating probable outcomes became standard practice for generations of physicists.[114]

Because of the insight and precision offered by quantum electrodynamics, Feynman called it "the jewel in the crown of quantum mechanics."

The Electro-Weak Theory—Quantum Mechanics Applied to the Weak Force

The description of the "weak force" particles *W* and *Z* (presented earlier, responsible for one process of the decay of atomic nuclei) was integrated around 1970 with quantum electrodynamics into what became known as the *electroweak theory*. It is this theory that explains the beta decay mentioned earlier and the bonding together of protons and neutrons in the nucleus of the atom, radioactivity, and why many of the heavier elements are unstable. Sheldon Glahow, Abdus Salam, and Steven Weinberg received the 1979 Nobel Prize in Physics *"for their contributions to the theory of the unified weak and electromagnetic interaction between elementary particles, including, inter alia, the prediction of the weak neutral current."* Their work was furthered by Gerrard 't Hooft, who would in 1999 share the Nobel Prize in Physics with his thesis advisor Martinus J. G. Veltman *"for elucidating the quantum structure of the electroweak interactions."*

Quantum Chromodynamics—Quantum Mechanics Applied to Color Charge and the Strong Force

As described earlier, this theory was developed by Murray Gell-Mann in the 1960s to describe the strong force that binds together quarks to produce the proton and neutron constituents of the atomic nucleus, in this case through the exchange of massless particles called gluons. In recognition of this work, Gell-Mann in 1969 received the Nobel Prize in Physics, *"for his contributions and discoveries concerning the classification of elementary particles and their interactions."*

The Higgs Field and Its "Carrier" Boson

While Gell-Mann was pondering the makeup of the neutron and the proton, Peter Higgs pondered where mass came from. As described earlier, in 1964 he and others proposed that the universe was filled with a third type of field, in addition to the fields of electromagnetism and gravitation.[115] Most of the particles of the Standard Model would acquire their masses through their interaction with this Higgs field. (As noted earlier, other individuals and groups had similar theories and ideas.)

As noted above, fields of force have associated "carrier particles" labeled as "gauge bosons."[116] In a similar manner, the Higgs field would suggest a new carrier particle or particles, to be called Higgs bosons. The July 2012 finding of a Higgs boson is strong evidence confirming the Higgs field as part of the Standard Model. More studies are needed to further verify that this Higgs boson behaves in accordance with the Standard Model, and maybe that there is more than one type of Higgs boson, as predicted by other theories. However, the Higgs boson was sufficiently established that the development of the theory would be recognized in 2013 through the award of the Nobel Prize in Physics to Peter Higgs and François Englert *"for the theoretical discovery of a mechanism that contributes to our understanding of the origin of mass of subatomic particles, and which recently was confirmed through the discovery of the predicted fundamental particle, by the ATLAS and CMS experiments at CERN's collider."* The latter's partner, Robert Brout, had died in the interim, or he too may have shared the prize.

D. THE CLASH OF RELATIVITY AND QUANTUM MECHANICS

In the course of our tour, we've followed the "map" provided by the very successful big bang model. At the core of this model is Einstein's general relativity, which is particularly successful in dealing with gravity, space-time, and the huge masses of the larger constituents of the cosmos. Later, we entered the realm of the Standard Model to explain the fundamental particles of matter and the forces of nature. This model has at its core quantum mechanics, which we know primarily as dealing with the very small. (Though actually it applies broadly.)

Quantum mechanics and relativity come together in two specific areas. First, as noted in the beginning of this chapter, gravity results from the curvature of space-time. But we talk in terms of a "gravitational field." One might expect then that quantum mechanics could be applied (in the manner described above for the fields and the forces of nature) to predict and describe the existence of a carrier particle called the "graviton" (which has been suggested but not found). And, second, at the big bang and at the center of black holes we have the combination of very high density of mass and a very small size. Relativity applies to the density of mass, and quantum mechanics applies for small size. The logical question then is, might there be some "theory of both"? Would it resolve the mathematical singularities at the point of the big bang and the core of a black hole? Would it describe what actually happens there? Is there, to use Hawking's words, "a theory of everything"? The answer, for now at least, is no, not yet. There seem to be some deep-seated conceptual and mathematical incompatibilities.

Two general approaches seem to be followed in pursuit of a merger of theories toward answers of these sorts of questions. One is called *loop quantum gravity*; the other, *string theory*. Both may be considered speculative, though considerable effort has been invested in each. And both are far beyond the expertise of this author and the scope of this book.

So I end this tour now. But I would like to point the way toward your exploration of these topics. For each approach I recommend an interesting book and provide you with an introduction to the subject: In regard to *loop quantum gravity*, I recommend the book that I cited several times at the beginning of this chapter: *Once before Time: A Whole Story of the Universe*, by Martin Bojowald (Reference FF). In regard to *string theory*, I recommend *The Little Book of String Theory*, by Steven S. Gubser (Reference II).

LOOP QUANTUM GRAVITY

Loop quantum gravity (LQG) applies the ideas of quantum mechanics toward describing the universe, including gravity. Since gravity results from distortions in space-time, LQG is a theory of quantum space-time. It attempts to treat gravity as a field of force in the same sense that quantum mechanics successfully treated electromagnetism in quantum electrodynamics.

In LQG, space takes on a granularity in the graviton in the same sense that photons represent the quantized granularity of electromagnetic fields. (Remember Section IV, earlier.) Space takes on the nature of finite loops, on the scale of the Planck length, 10^{-35} meters, woven into a network, or fabric. There is thus a smallest size to the elements of space.

This theory is being pursued in two general directions: spinfoam theory (covariant LQG) and canonical LQG. One consequence of the theory is that the evolution of the universe may be traced back to times before the big bang.

STRING THEORY

String theory has been an approach toward describing physical phenomena dating back at least to the 1960s, when it was used in an attempt to model the strong nuclear force. It was subsequently used to describe only bosons. Later, superstring theory was developed to show a connection between bosons and fermions in what is called *supersymmetry* (also mentioned in Section IV). In the mid-1990s, five versions of the theory were combined as parts of an eleven-dimension M theory; and in 1997, it was discovered that string theory may relate to quantum field theory.

Instead of particles, the fundamental elements of nature are minuscule, one-dimensional "strings." Its properties are determined by its vibrational states, which give it charge, mass, and other properties. But the overall theory does not apply universally; it's more that the individual five theories work in an overlapping fashion, each applying perhaps with one neighboring theory to a particular area of physics. String theory also describes a multitude of possible universes. In that regard, I would suggest again that you read Greene's *The Hidden Reality—Parallel Universes and the Deep Laws of the Cosmos* (Reference AA). K. C. Cole in *Quanta Magazine*, September 15, 2016, describes how string theory has so far failed to live up to its promise of uniting gravity and quantum mechanics, but also suggests that the theory has produced significant advances in mathematics and other aspects of physics.[117]

In this book, *Quantum Fuzz,* we next turn to the practical results of quantum mechanics, for me its greatest gift. I show in Part Four how, based on a few simple concepts, quantum mechanics explains chemistry and the atoms of all of the elements, the building blocks of everything around us. In Part Five I describe the many inventions and practical products that are explained by or have directly resulted from our understanding of this theory, including levitated trains, atomic monolayer materials, MRI diagnostic medical imaging, and potential sources of unlimited power for the future.

Part Four

THE MANY-ELECTRON ATOM AND THE FOUNDATIONS OF CHEMISTRY AND MATERIALS SCIENCE

"THE PERIODIC TABLE."

Chapter 10

INTRODUCTION TO PART FOUR

In Part Four we address something especially beautiful in its inherent simplicity,[1] the understanding of the atom, the foundations of chemistry, and the makeup of everything that we see around us. Here we'll see not only how the properties of the elements are explained by quantum mechanics but also how the strange cycle of their properties is related to the electronic structure of the atoms, and how that determines their bonding to form most of what we see around us and what may be created. (For me this is the most important product of quantum mechanics: the provision of an understanding that is the practical engine of invention.)

APPROACH

Here we extend our understanding of the states of the electron in hydrogen to show what may be expected for the atoms of the rest of the elements. Our approach will be somewhat empirical and in no way presumes the full "reduction" that concerns Scerri.[2] However, as will be explained more fully in Chapter 14, a lot of the physics is in place, and quantum mechanics is rather compelling in providing much of our understanding of the elements and the periodic table.

I begin, in Chapters 11 and 12, by more completely defining the properties of the electron. In Chapter 13 I explain the consequences of exclusion in determining the properties of the elements and the cyclic nature of those properties as displayed in the periodic table. In Chapter 14 I examine what physically is going on in the atom to produce its chemistry and determine its size. (Strange as this may seem, the sizes of atoms are actually smaller and smaller as we consider successively heavier and heavier elements with more and more electrons within each period of the periodic table.) In Chapters 15, 16, and 17, I show how the chemical properties of the elements result in bonding to form molecules and solid materials that insulate, conduct, or semiconduct (to produce all of the modern electronics and the other devices described in Part Five).

At times I may operate conceptually beyond the level of most introductory chemistry and physics college courses, but I leave out the math normally involved and I use methods requiring no special training in math or science. Should you find the discussion at any point to be more detailed than you would like, I would suggest that you scan through indicated sections to get a sense of the physics involved, recognizing that this Part Four also provides a valuable background for understanding the fascinating inventions, the "quantum wonders," yet to be described in Part Five.

THE PERIODIC TABLE

Everything around us is made up of the atoms or combinations of the atoms of the elements. As of October 13, 2016, 118 different types of atoms have been discovered or created.[3] These are the building blocks, respectively, of the 118 different elements, each with its own distinct set of properties. As we consider each successively heavier element, one after the other, the properties of these atoms, these elements, seem to cycle periodically, nearly repeating themselves. I summarize this behavior by listing the elements in what is called a *periodic table*. A brief, interesting history of the development of such tables, and a bit of the life of the charismatic character mainly responsible for that development is presented in Appendix B. I would recommend that you take a break from the theory at this point, and just read through Appendix B for the entertainment and background that it provides.

Many different forms of periodic tables have been produced: some listing the elements by atomic number in columns, some in spirals, some in rows. What is common to all of them is that each period in which the properties of the elements seem to repeat themselves is marked by the presence of one of the "inert" elements that were originally thought to react and combine with no other elements, that is, one of the noble gasses. And the numbers of elements between each pair of noble gasses is always the same, regardless of the form of the table.

In Table B.2, for example, counting through the elements by atomic number successively, starting with hydrogen, atomic number 1, at the bottom left, we immediately have the inert noble gas helium, atomic number 2, at the bottom right. We move **2** elements to helium. Through the next row we have **8** more elements to the noble gas neon. Through the next row, we have **8** more to argon, then **18** more to krypton, **18** more to xenon, and **32** more to radon. This "**2, 8, 8, 18, 18, 32**" sequence will be the same in all periodic tables.

THE ELECTRONIC STRUCTURE OF THE ATOM

Scerri provides an excellent history of how the electron structure of the atom was determined in his Chapter 7, "The Electron and Chemical Periodicity" and in his Chapter 8, "Electronic Explanations of the Periodic System Developed by Chemists."[4] I summarize this history very briefly now as part of this introduction.

With the discovery of the electron in 1897 and the subsequent realization that the chemical properties of the elements are related to the number of electrons in their atoms, physicists and chemists sought to understand the periodic table in terms of the population of electron states by electrons in the atoms of each of the elements. The chemists were inductive and empirical, and they focused mainly on the manner in which the elements interacted with other elements. They were fairly successful in evolving an empirical set of rules that described the arrangement of the table.

The physicists, meanwhile, sought to overcome some of the contradictions of classical theory and began in 1900 to evolve a quantum theory (as we've described) that might allow the periodicity of the table to be deduced from the physics of the atom. As presented earlier, the lead proponent in this activity was the Danish physicist and later Nobel Prize winner Niels Bohr. But, as Scerri points out, Bohr's approach was not so much predictive as empirical in guiding the form of the physics based on what he already knew about the elements and the table.

The physicists succeeded in solving exactly for the possible spatial states for the electron in the hydrogen atom, as described in the chapters of Part Two of this book. While that success has been difficult to extend mathematically to many-electron atoms except by approximation, the hydrogen results nevertheless provide a guide for qualitatively predicting the electronic structure of these more complex atoms and for understanding the arrangement of the periodic table. As you will see, it all depends on energy.

Chapter 11

ENERGY, MOMENTUM, AND THE SPATIAL STATES OF THE ELECTRON IN THE HYDROGEN ATOM

Note: In the indented paragraphs of this chapter I describe some important fundamental concepts that will help in understanding the physics of the atom. And I also use a few numbers here and there in these indented sections to provide a sense of magnitudes. These concepts are central to all of this Part Four, so it's worth spending a little extra time here to take in what is presented.

IT ALL DEPENDS ON ENERGY

Remember the various spatial-state solutions to Schrödinger's equation for the hydrogen atom that were described in Chapter 3. Well, one might infer from the pronounced differences in the sizes of spatial-state cloud cross sections shown in Figure 3.8 that there are also pronounced differences in the energy levels of these states. And that is true. Schrödinger calculated these energy differences in the process of solving for the spatial states. But we need to *understand why* his calculations got these results. And before that we need to define what *energy* means in the context of the planets and the atom, as follows.

Objects that move around freely have a *positive kinetic* energy of motion depending on how heavy they are and how fast they are moving. In contrast, an object under some force of attraction is defined to have a *negative potential* energy.

The potential energy is made more negative if the object is placed closer to the source of attraction where the attraction is stronger. The object could be a planet attracted by the gravity of the sun or an electron attracted by the pull of its negative charge toward the positive charges of the protons in an atomic nucleus.

If an object is in some sense moving (i.e., has kinetic energy) in the presence of an attraction, then it can break free of the attraction if its positive kinetic energy is larger than its negative potential energy so that the sum of the two energies is greater than zero. If this sum is less than zero, negative (that is, if the negative potential energy of attraction overcomes the positive kinetic energy of motion), then the object will be in a *bound* state, perhaps still moving around in some sense but unable to break completely free of the attraction. The more negative this *total* energy, the more tightly the object is bound.

So, it's quite in keeping with the forces of nature that Schrödinger found that the spatial states of lower total energy would also more tightly bind the electron to the nucleus and have smaller probability-cloud cross sections. (Remember, though, what is special about his solutions is that they allow only certain discrete bound states, that is, states having only certain discrete quantized energies, not the continuum of energies and orbit sizes that would be available from classical physics.)

The potential energies of the planets in the solar system are very large negative numbers, large enough to overcome the very large positive kinetic energies of these rapidly moving huge masses and to bind them in their orbits. In contrast, the energies of the very small and light electrons in the atom are quite small. For the hydrogen atom in particular, the energies of the bound states of the electrons are small enough so that we could (in principle) remove hydrogen's one electron from any of its bound states using a couple of nine-volt batteries.

We are now going to demonstrate this using a thought experiment. Imagine that we hook the two batteries together in series (negative terminal of one to the positive terminal of the other) so that with the two of them we have a total of eighteen volts. Scientists call that an "electric potential" of eighteen volts. Now if we could attract a single electron to move from the unconnected negative terminal of one of the batteries to the positive terminal of the other battery (at its electric potential of eighteen volts), we could impart an energy to the electron of eighteen electron volts; in scientific shorthand, that is 18 eV. If we moved two electrons, we would provide an energy of 36 eV. It's that simple. (The *e* in "eV" refers to the charge on the electron. When a charge is moved by an electric potential, whether it's the charge on an electron or something else, it acquires energy that can be measured in electron volts.)

The solution to Schrödinger's equation gives −13.60 eV as the $n = 1$ energy level of the electron in hydrogen's fundamental most tightly bound, 1s state, its ground state (probability cloud shown at the bottom left of Fig. 3.8). With our batteries, we could provide an 18eV energy boost that would not only lift the electron free of the hydrogen atom but also leave it with an extra 4.4 eV of kinetic energy so that it could speed off to somewhere else.

Now let's examine the energy levels in the rest of the spatial states of the electron in the hydrogen atom. These include the bound spatial states, some of which are represented by the probability-cloud cross sections displayed in Figure 3.8.

As part of the solutions to his equation, Schrödinger found that each bound state for the electron in hydrogen has a total negative energy at only one of an infinite number of possible discrete allowed energy levels characterized by the so-called *primary quantum numbers* labeled generally by the symbol n. His solutions provide that n can only equal the integers 1, 2, 3, or 4, and so on. These are the numbers shown before the letters above the probability-cloud cross section representations for some of the individual states in Figure 3.8. (They are also the energy quantum numbers for the atomic orbits of the Bohr model constructed in the early days of the development of the theory.) Note that there can be more than one state at each energy level, as will be discussed a bit further on in this chapter.

The solved-for energy of each energy level is −13.60 eV *divided by* n^2, so at each successive energy level the energy gets to be less negative. (That is because *negative* 13.60 eV is being divided by an increasingly larger squared integer.) So, calculating the energies for the first seven levels, we find the following, starting with the highest level (having the smallest negative energy) with states that least tightly bind the electron. For energy level:

$n = 7$ the energy of every state is $(-13.60)/(7\times7) = -0.28$ eV
$n = 6$ the energy of every state is $(-13.60)/(6\times6) = -0.38$ eV
$n = 5$ the energy of every state is $(-13.60)/(5\times5) = -0.54$ eV
$n = 4$ the energy of every state is $(-13.60)/(4\times4) = -0.85$ eV
$n = 3$ the energy of every state is $(-13.60)/(3\times3) = -1.51$ eV
$n = 2$ the energy of every state is $(-13.60)/(2\times2) = -3.40$ eV
$n = 1$ the energy of every state is $(-13.60)/(1\times1) = -13.60$ eV

You can see the trend: the higher and higher the n value, the closer and closer to zero is the negative energy level, at which point the electron becomes free and is not bound at all. States with very high n have very small negative energies, and electrons in these states are therefore not very tightly bound and are correspondingly relatively large in size. The trend in size as n increases is illustrated in Figure 3.8.

AND ANGULAR MOMENTUM

Another property of each state that results from the solution of Schrödinger's equation is the state's angular momentum. Angular momentum tends to keep a rotating object rotating or moving in a circular or elliptical motion, just as linear momentum tends to keep an object moving at its same speed and in its same direction. If an object rotates one way, the angular momentum is said to be positive; if it rotates the other way, the angular momentum is said to be negative.

While Schrödinger's probability-cloud cross sections don't rotate or spin, the states that they represent nevertheless have *spatial-state angular momentum* that is indicated by the letters above the cross sections. (There are units for angular momentum, just as there are units like eV for energy. And nature's basic unit of angular momentum is [you guessed it] Planck's constant divided by 2π, in scientific shorthand $h/2\pi$.)

Schrödinger's results tell us that, like the state's energy, the magnitude[1] of its spatial-state angular momentum is quantized, as noted by the angular-momentum quantum numbers labeled generally by the letter l, where l can equal only the integers 0, 1, 2, or 3, and so on, providing that l is always less than n. Spectral lines resulting from states having these l numbers are found to have *sharp*, *principal*, *diffuse*, and *fundamental* characteristics found in spectra, so the spatial-state angular-momentum quantum numbers $l = 0, 1, 2,$ or 3 have historically come to be represented, respectively, by the letters s, p, d, or f.

And so it is that the clouds' cross section representations shown in the first "column" of Figure 3.8 for the spatial states with $l = 0$ and energies indicated by $n = 1$ and $n = 2$ are labeled 1s and 2s, respectively. And the cloud cross sections shown in the second "column" of Figure 3.8, each with $l = 1$ but, respectively with $n = 2$ and $n = 3$, are labeled 2p and 3p, while the single-cloud cross section of the third "column," with $n = 3$ and $l = 2$, is labeled 3d.

Now we can understand an interesting result observed in hydrogen spectra. According to quantum mechanics, a photon has either +1 or −1 units of angular momentum, depending on the polarization of the photon that carries away the energy given off in the transition. (Strange "beasts" these photons: as determined by quantum mechanics they all have only this plus or minus one unit of angular momentum, even though they have no mass [i.e., weight in a gravitational field]! And despite this they can have widely different energies, depending on the energy levels of the states involved in the transition that produces them!)

So, for the transition of an electron from an excited state in a hydrogen atom, the angular momentum of the excited state and the angular momentum of the lower energy state that the electron transitions into must differ by the one unit of angular momentum that is carried away by the photon emitted in the transition. Just as energy is conserved in a transition, and the energy of the initial state equals the energy of the final state plus the energy of the photon; so angular momentum is also conserved, so that the angular momentum of the initial state equals the angular momentum of the final state plus the angular momentum of the photon. This means that s states can transition only into p states, p states can transition only into s or d states, d states can transition only into p or f states, and so on. There are thus fewer lines predicted for the hydrogen spectrum than otherwise would have been expected, and this is precisely what has been observed. And the same holds true for the transitions of the electron between states in the rest of the elements.

Angular momentum can be symbolized by an arrow (a *vector*) with length in proportion to the angular momentum's magnitude and pointed in a direction along an axis of rotation. Such a vector can be thought of as representing angular momentum for each of the spatial states. (Yes, it's strange, there is angular momentum but no evidence of orbit or rotation.) This vector can be broken down into two parts, that is, two *components*: one representing the portion of the vector that might lie in the direction of a magnetic field (if a magnetic field were applied) and the second representing the portion that would then lie at a right (90-degree) angle to the magnetic field. (An atom might find itself in the earth's magnetic field or in an experiment between the "jaws" of a horseshoe magnet. Even if the electron is not in a magnetic field, its possible components of angular momentum are defined by what would happen if it were in a magnetic field.) The component of the angular momentum that would line up in the direction of a magnetic field is also quantized as a part of each spatial-state solution to Schrödinger's equation. This component is represented by a particular integer (which we refer to generally as *m*) times one basic unit of angular momentum: that is, $m \times h/2\pi$. And *m* can have any integral value in the limited range from $-l$ to $+l$.[2] For reasons that will be explained in Chapter 12, *m* is known as the *magnetic* quantum number.

Note that *m* values are also shown just above the cross sections of the probability clouds in Figure 3.8. The +1 or −1 values that are shown for several of the cross sections indicate that these cloud cross sections may represent either of two spatial states, one with $m = +1$ or one with $m = -1$, where +1 is for a state with the cross section shown and −1 is for a state that has a similar cross section that lies in a plane that is perpendicular to the plane of the page.

So, we've defined the three properties inherent in Schrödinger's *spatial*-state solutions for the hydrogen atom. Every spatial state for the electron is characterized by its own set of these properties, labeled for each state by the quantum numbers *n*, *l* and *m*, plus one more: its intrinsic angular momentum, otherwise simply called *spin*, described as follows in Chapter 12.

Chapter 12

SPIN AND MAGNETISM

SPIN, A FUNDAMENTAL PROPERTY OF THE ELECTRON

As described in Chapter 2, Pauli *postulated* the existence of spin to help Bohr explain the properties of the elements and the periodic table. It worked, but there was no firm scientific basis. That basis was put in place in one magnificent theoretical work by the mathematician and physicist Paul Dirac in 1928.[1] Using Einstein's special relativity and only the electron's electrical charge and mass, he calculated *all* of the rest of the properties of the isolated electron (that is, the *intrinsic* properties of the electron, regardless of whether or not it is in an atom).[2] (Remember that it was Dirac, along with Heisenberg and Schrödinger, each using a different mathematical approach to quantum mechanics, who three years earlier had all successfully calculated the energies and spatial states for the hydrogen atom, and all got the same results, though Schrödinger's format was more amenable to interpretation.)

The properties that Dirac derived included a very tiny *intrinsic* angular momentum that would be present even if the electron were completely isolated from the atom and regardless of the overall shape of the spatial state that the electron might take on. Even though there is nothing in what Dirac did to indicate that the electron is in any way spinning, the term used to describe this intrinsic angular momentum was (and still is) *spin* (as coined earlier by Uhlenbeck and Gaudsmit). Dirac calculated that the spin of an electron can have only two values, either +½ or −½ times one basic unit of angular momentum, which, as described earlier, is Planck's constant divided by 2π, written $h/2\pi$.[3] And so we see that spin is also *quantized*, with just two possible values, that is, two possible spin quantum numbers. We refer to these here simply as "plus spin" and "minus spin." (It is because of this binary spin state, that experiments with the photon [with its binary angular-momentum polarization states, as described in Chapter 11] can be straightforwardly substituted for spin in experiments, as was done in the confirmation of Bell's inequality, as described in Chapter 6.)

MAGNETISM

If a bar magnet is placed in a magnetic field (for example, between the jaws of a horseshoe magnet) it will be attracted into the strongest part of the field, and it will line up to point toward a magnetic field's north pole, just as a compass needle points toward the earth's north pole. We say that the bar magnet has a *magnetic moment*.

If an isolated electron is placed in a magnetic field, it behaves in a similar way to the bar magnet, demonstrating that it too has a magnetic moment. Because the electron always has this magnetic moment regardless of where it happens to be, we say that it has an *intrinsic magnetic moment*. And quantum mechanics shows that the electron's intrinsic magnetic moment has a magnitude and direction in proportion to its spin angular momentum.[4] If the electron is in a plus spin state, then its magnetic moment aligns with the magnetic field as described, and if it is in a minus spin state, it aligns in the opposite direction.

Remember from Chapter 11 that the component of the *spatial-state* angular momentum that will line up in the direction of a magnetic field is quantized as a part of each spatial-state solution to Schrödinger's equation. This alignment occurs because the electron also has a *spatial-state magnetic moment* in proportion to this same component of spatial-state angular momentum. And remember that this component is quantized and represented by an integer that we refer to generally as m. (So we now understand why m is referred to as the "*spatial-state magnetic quantum number*.")

It is the combination of intrinsic spin and m that determines the magnetic properties of the electron for each combined spin and spatial state that it may be in (occupy). There is a very small quantized shift in the energy for each spin and spatial-state combination (thinking classically for the moment), depending on how the tiny "bar magnets" of the electron's intrinsic magnetic moment and spatial-state magnetic moment line up with each other to attract or repel. This "fine-structure splitting" is so small that we can ignore it here, but each energy shift has been measured and verified to occur exactly according to theory.

Chapter 13

EXCLUSION AND THE PERIODIC TABLE

Now an understanding of the elements, their various properties, essentially everything that we see around us, is within our grasp! We use the states of the electron in hydrogen as a guide in a progression through four related tables. These are *unusual* tables. They don't contain boring data. Their contents are broken down into sections that have physical meaning. And one table flows to the next until (*voilà*) *in the fourth table we have a modern periodic table* of the elements! And we have achieved this through a progression that simply reveals the atomic structure of the elements as we progress easily from what we already know about the hydrogen atom.

Before proceeding with this chapter, it's best to get an overview first. For this I ask you to just scan to get a "snapshot" of each of the tables (to facilitate this scan, and for easy reference, I would suggest—if it is convenient for you to do so—that you make a separate copy of each of the tables.) When you have finished this scan, glance at the short notes about each table on the page that immediately follows them. This will be like fanning through a succession of slightly changed cartoon pages to get the effect of a motion-picture show.

Table I. Properties of 128 of the Lowest Energy Combined Spin

Each square marks a state, with energy, angular momentum, and spatial and spin
and column in which it is located. States for energy levels n = 4 and n = 5 of the

Unlike those shown in Figure 3.8, the probability cloud cross sections representing some of the states here do not show relative size. (They have been taken from Figure 5-5 of Leighton, Reference F, with permission from Margaret L. Leighton.)

Energy Level, n	Plus Spin			
	m =3	m =2	m =1	m =0
5				
4				

Energy Level, n	s States, $\ell = 0$	
	Plus Spin	Neg. Spin
	m =0	m =0
7	state	state
6	state	state
5	state	state
4	state	state
3	state	state
2		
1		

d States, $\ell = 2$						
Plus Spin						
m =2	m =1	m =0	m =-1	m =-2	m =2	m =1
state	state	state	state	state	state	state
state	state	state	state	state	state	state
state	state	state	state	state	state	state
			state	state		
			state	state		

and Spatial States of the Electron in the Hydrogen Atom.

magnetic moments as indicated by the quantum numbers of the row
f block of states are shown separately at the top of this table.

f States, $\ell = 3$										
			Minus Spin							Row
m =-1	m =-2	m =-3	m =3	m =2	m =1	m =0	m =-1	m =-2	m =-3	
state	state	state	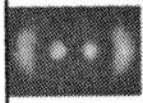		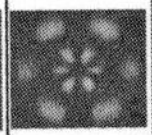		state	state	state	5
state	state	state		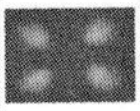			state	state	state	4

Minus Spin		
m =0	m =-1	m =-2
state	state	state
state	state	state
state	state	state
	state	state
	state	state

p States, $\ell = 1$						
Plus Spin			Minus Spin			Row
m =1	m =0	m =-1	m =1	m =0	m =-1	
state	state	state	state	state	state	7
state	state	state	state	state	state	6
state	state	state	state	state	state	5
state	state	state	state	state	state	4
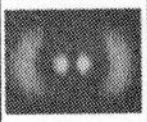		state			state	3
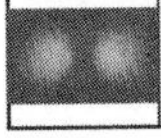		state	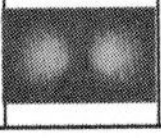	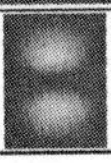	state	2
						1

Table II. Properties of 128 of the Lowest Energy States

**Insert this entire row into Row #7 after the **state below.
*Insert this entire row into Row #6 after the *state below.

	Plus Spin			
	m =3	*m* =2	*m* =1	*m* =0
5f	state	state	state	state
4f	state	state	state	state

	s States, $\ell = 0$	
	Plus Spin	Neg. Spin
	m =0	*m* =0
7s	state	state
6s	state	state
5s	state	state
4s	state	state
3s	state	state
2s	state	state
1s	state	state

	d States, $\ell = 2$						
	Plus Spin						
	m =2	*m* =1	*m* =0	*m* =-1	*m* =-2	*m* =2	*m* =1
7d	state	state	state	state	state	state	state
6d	state **	state	state	state	state	state	state
5d	state *	state	state	state	state	state	state
4d	state	state	state	state	state	state	state
3d	state	state	state	state	state	state	state

The labels to the left of the blocks correspond to the original energy level and angular momentum of the comparable states for the electron in the hydrogen atom, as displayed in Table I.

for the Electron in a Generic Many-Electron Atom.

f States, $\ell = 3$										
			Minus Spin							Row
m =-1	*m* =-2	*m* =-3	*m* =3	*m* =2	*m* =1	*m* =0	*m* =-1	*m* =-2	*m* =-3	
state	state	state	state	state	state	state	state	state	state	7
state	state	state	state	state	state	state	state	state	state	6

Minus Spin		
m =0	*m* =-1	*m* =-2
state	state	state
state	state	state
state	state	state
state	state	state
state	state	state

	p States, $\ell = 1$						
	Plus Spin			Minus Spin			Row
	m =1	*m* =0	*m* =-1	*m* =1	*m* =0	*m* =-1	
7p	state	state	state	state	state	state	7
6p	state	state	state	state	state	state	6
5p	state	state	state	state	state	state	5
4p	state	state	state	state	state	state	4
3p	state	state	state	state	state	state	3
2p	state	state	state	state	state	state	2

Table III. Expected Outermost Occupied Electron State for the

**Insert this entire row into Row #7 after Element #89 ** below.

*Insert this entire row into Row #6 after Element #57 * below.

	Plus Spin			
	m =3	m =2	m =1	m =0
5f	90	91	92	93
4f	58	59	60	61

	s States, $\ell = 0$	
	Plus Spin	Neg. Spin
	m =0	m =0
7s	87	88
6s	55	56
5s	37	38
4s	19	20
3s	11	12
2s	3	4
1s	1	2

	d States, $\ell = 2$						
	Plus Spin						
	m =2	m =1	m =0	m =-1	m =-2	m =2	m =1
7d							
6d	89 **						
5d	57 *	72	73	74	75	76	77
4d	39	40	41	42	43	44	45
3d	21	22	23	24	25	26	27

The labels to the left of the blocks correspond to the original energy level and angular momentum of the comparable states for the electron in the hydrogen atom, as displayed in Table I.

Atom of Each Element, as Marked by Its Atomic Number.

f States, $\ell = 3$										
			Minus Spin							Row
m =-1	m =-2	m =-3	m =3	m =2	m =1	m =0	m =-1	m =-2	m =-3	
94	95	96	97	98	99	100	101	102	103	7
62	63	64	65	66	67	68	69	70	71	6

Minus Spin		
m =0	m =-1	m =-2
78	79	80
46	47	48
28	29	30

	p States, $\ell = 1$						
	Plus Spin			Minus Spin			Row
	m =1	m =0	m =-1	m =1	m =0	m =-1	
7p							7
6p	81	82	83	84	85	86	6
5p	49	50	51	52	53	54	5
4p	31	32	33	34	35	36	4
3p	13	14	15	16	17	18	3
2p	5	6	7	8	9	10	2
							1

Table IV. A Modern Arrangement of the Periodic

** Actinide Series (Insert the entire row after Ac, Z = 89 below)

90 **Th** Thorium	91 **Pa** Protactinium	92 **U** Uranium	93 **Np** Neptunium	94 **Pu** Plutonium	95 **Am** Americium

** Lanthanide Series (Insert the entire row after La, Z = 57 below)

58 **Ce** Cerium	59 **Pr** Praseodymiu	60 **Nd** Neodymium	61 **Pm** Promethium	62 **Sm** Samarium	63 **Eu** Europium

IA	IIA	Group	IIIB	IVB	VB	VIB	VIIB	-----	VIIIB
87 **Fr** Francium	88 **Ra** Radium		89 **Ac**** Actinium						
55 **Cs** Cesium	56 **Ba** Barium		57 **La*** Lanthanum	72 **Hf** Hafnium	73 **Ta** Tantalum	74 **W** Tungsten	75 **Re** Rhenium	76 **Os** Osmium	77 **Ir** Iridium
37 **Rb** Rubidium	38 **Sr** Strontium		39 **Y** Ytterium	40 **Zr** Zirconium	41 **Nb** Niobium	42 **Mo** Molybdenum	43 **Tc** Technetium	44 **Ru** Ruthenium	45 **Rh** Rhodium
19 **K** Potassium	20 **Ca** Calcium		21 **Sc** Scandium	22 **Ti** Titanium	23 **V** Vanadium	24 **Cr** Chromium	25 **Mn** Manganese	26 **Fe** Iron	27 **Co** Cobalt
11 **Na** Sodium	12 **Mg** Magnesium								
3 **Li** Lithium	4 **Be** Beryllium								
1 **H** Hydrogen									

Metals (white box)

Metalloids = Semiconductors (shaded box)

Table of 103 Elements (Same as Table B.2).

								Row
96 **Cm** Curium	97 **Bk** Berkelium	98 **Cf** Californium	99 **Es** Einsteinium	100 **Fm** Fermium	101 **Md** Mendelevium	102 **No** Nobelium	103 **Lr** Lawrencium	7
64 **Gd** Gadolinium	65 **Tb** Terbium	66 **Dy** Dysprosium	67 **Ho** Holmium	68 **Er** Erbium	69 **Tm** Thulium	70 **Yb** Ytterbium	71 **Lu** Lutetium	6

	IB	IIB	IIIA	IVA	VA	VIA	VIIA	VIIIA	Row
									7
78 **Pt** Platinum	79 **Au** Gold	80 **Hg** Mercury	81 **Tl** Thallium	82 **Pb** Lead	83 **Bi** Bismuth	84 **Po** Polonium	85 **At** Astatine	86 **Rn** Radon	6
46 **Pd** Palladium	47 **Ag** Silver	48 **Cd** Cadmium	49 **In** Indium	50 **Sn** Tin	51 **Sb** Antimony	52 **Te** Tellurium	53 **I** Iodine	54 **Xe** Xenon	5
28 **Ni** Nickel	29 **Cu** Copper	30 **Zn** Zinc	31 **Ga** Gallium	32 **Ge** Germanium	33 **As** Arsenic	34 **Se** Selenium	35 **Br** Bromine	36 **Kr** Krypton	4
			13 **Al** Aluminum	14 **Si** Silicon	15 **P** Phosphorous	16 **S** Sulfur	17 **Cl** Chlorine	18 **Ar** Argon	3
			5 **B** Boron	6 **C** Carbon	7 **N** Nitrogen	8 **O** Oxygen	9 **F** Fluorine	10 **Ne** Neon	2
								2 **He** Helium	1

Nonmetals

Table I is a set of squares, each of which represents one of the 128 lowest-energy states of the hydrogen atom—starting with the states of lowest energy in the bottom row. The various rows, columns, and four groups organize these states according to their energy, angular momentum, and spin properties. The cloud cross sections for forty of these states are shown in some of the boxes.

Table II represents the tables of states of all of the rest of the elements. Each will have its own table, but the tables are somewhat similar, and so we represent them all with this one *generic table.* No cloud cross sections are shown, because these will be somewhat different for the atoms of each element, and only somewhat changed from those shown for hydrogen in Table I. For reasons that will be described, the d block of states in the middle is raised upward by one row to higher energy, and the f block is shifted upward by two rows, as compared to the position of these blocks in Table I.

Table III just numbers the states of Table II from left to right across each row—starting with the bottom row. Each number marks the last state that would be occupied in a table of states for an atom with that atomic number, that number of electrons. This assumes that the electrons of that atom will successively occupy the lowest energy states, one electron per state, starting with the state at the left end of the first row, and then successively filling states to the right in that row and then up to the left end of the next row, and so on.

Table IV is just Table III with the name of each element added into the square (now stretched to a rectangle) where its atomic number appears—and with Roman-numeral "group" labels for the columns inserted above the lower blocks of elements. And, as noted above, *voilà!* Now we have a modern periodic table of the elements!

If you were to flip Table IV bottom to top, it would take on a more familiar form, with the lightest elements shown in the first row and the heaviest elements at the bottom. And remember, Table III gives us the spin and spatial state of the outermost electron for each of the elements shown in Table IV. We'll see how that state affects the properties of each element.

In the rest of this chapter I'll describe more fully the three steps from

Table I for the hydrogen atom to the periodic table (Table IV). And we'll see how Table III reflects the degree to which the p states at each energy level are occupied and how, for every element, this determines its chemical properties. Then in Chapter 14 we'll see what is happening inside the atom as we consider heavier and heavier elements with more and more electrons. First let's consider our tables, and then I'll describe a bit more about the progression to a periodic table that I have just outlined.

THE HYDROGEN ATOM AS A GUIDE TO THE ATOMS OF THE REST OF THE ELEMENTS

We start with Table I, a concise summary of the properties of 128 of the infinity of states available to the electron in the hydrogen atom. These 128 are the states with the lowest levels of energy and angular momentum (those states most likely to be occupied by the electron, recognizing, again, that things in nature tend to occupy a lower- or the lowest-energy state). Table I has the following features:

1. Each square marked "state" represents one unique state that the electron may occupy. Each state has a unique set of properties noted by the four quantum numbers for that state.

   The states are grouped into four major blocks. Considering each block successively from left to right and then up: the first block contains those squares marking s-states, states with angular momentum indicated by the quantum number $l = 0$; the second block contains those squares marking d-states, states with angular momentum indicated by the quantum number $l = 2$; the third block contains those squares marking p-states, states with angular momentum indicated by the quantum number $l = 1$; and the fourth block above contains those squares marking f-states, states with angular momentum indicated by the quantum number $l = 3$.
2. Each block is divided into block halves, with the left half containing squares that mark those states with plus spin, and the right half marking those states with minus spin.

3. Each half of each block is divided into columns, with each column indicating a magnetic quantum number $m = 0$, 1 or -1, 2 or -2, and so on, of the states within that column; note that each m value represents the component of angular momentum that would align parallel (or antiparallel) to the direction of a magnetic field.
4. Finally, each row of the table (numbered at the right of the table) indicates the corresponding principal quantum number, $n = 1$, 2, 3, 4, and so on (shown at the left side of the table); note that each n represents the energy level of all states within that row.
5. To indicate some of the variety of forms of the probability clouds of the various states, forty of the "state" labels have been replaced by the probability-cloud cross sections for those states. (All of these cloud cross sections have been magnified differently so that they just fill the squares. The relative sizes of five of these cross sections, all under the same magnification, are displayed in Figure 3.8.

CLUES TO THE CHEMISTRY OF THE ELEMENTS IN A GENERIC TABLE OF STATES

Next consider generic Table II, which (in a manner similar to Table I for hydrogen) *represents* the states and properties available to electrons in atoms having more than one electron. The similarity between these two tables results mainly because Schrödinger's solutions for the states of the many-electron atom are much the same as those found for the hydrogen atom.

However, Table II differs from Table I in three ways. The first difference results from a difference in the physics and is of fundamental importance, as described immediately below.

1. Note that the entire d block of states in Table II has been raised one row (one energy level) higher than the d block in Table I, and the entire f block has been raised two rows (two energy levels) higher. These shifts result because of a physical interaction that is present in the many-electron atom but not present in the hydrogen atom. In the many-electron atom, each electron, wherever it may be in its

probability cloud, is screened from the attraction of the positively charged protons of the nucleus by the negative charge of each of the other electrons, to the degree that the probability clouds of these other electrons lie closer to the nucleus. So, those states that on average lie farther out and are more screened are attracted less, have higher (less negative) energies and are less tightly bound. Specifically, the d states are more screened and have less negative energies, and so the d block of states is higher in energy (and in the table) than the s and p blocks of the same n, and the f block of states is similarly higher in energy than the d block of states.

2. The second difference between the tables is in the labels to the left of each block of states in Table II. These labels correspond to what were the original energy level and angular momentum of each row of states before the shift, and to the comparable rows of states for the electron in the hydrogen atom. For example, the bottom row of the d block, now in the fourth row of Table II, is labeled as 3d, because these states are somewhat like the hydrogen states in the third row of the d block in Table I.
3. Finally, no probability-cloud cross sections are shown in Table II. That is because a particular state (say $n = 2$, $l = 1$, $m = 0$, plus spin) of the atom of one element is somewhat different from a state with the same quantum numbers in an atom of another element, and their cloud cross sections are somewhat different. Since this is a generic table, pertaining to all many-electron atoms, it would be inappropriate to show the cloud cross sections of the states of the atom of any particular element. We can get some idea of what these states may look like by examining the cross sections of comparable states for the electron in hydrogen, as shown in Table I.

So, we have all of these states with all of their associated properties. So what? Surprisingly (as you will soon come to see) *much of an understanding of the chemistry of each element comes from simply counting the number of states occupied by electrons at each energy level in its atoms.*

For starters, note that Row #1 of Table II has just 2 states in it; Row #2 has 8; Row #3 has 8; Row #4 has 18; Row #5 has 18; and Row #6 has

32, including those states in Row #6 of the f block of states at the top of the table. This "2, 8, 8, 18, 18, 32" sequence is *exactly* that found for the number of elements between "inert" elements as one counts through the elements in order of atomic weight or atomic number. And this correlation is not a coincidence, as will become clear as we proceed.)

ANTISOCIAL BEHAVIOR (EXCLUSION)

Now we discuss something that is fundamental to our understanding of the properties of the atoms of all of the elements after hydrogen, and fundamental to our understanding of the arrangement of the periodic table.

Remember that *exclusion* was originally postulated by Wolfgang Pauli, in what is now known as the *Pauli exclusion principle,* to help Bohr use his model of the atom to explain the elements and the periodic table. That was just a postulate. Since then, scientists have shown exclusion to be a fundamental property of electrons.

As it happens, particles have states described either by symmetrical or antisymmetrical wavefunctions. (Symmetry is in the mathematics, beyond the scope of explanation here. But the results of the math are profound!) Those particles having symmetrical wavefunctions all have integral or zero spin and are called bosons. (Remember the fundamental bosons of the Standard Model described in Chapter 9.) Bosons may all occupy the same state. Those particles having anti-symmetric wavefunctions all have half integral spin and are called fermions particles. (Again, remember the fundamental fermions described in Chapter 9.) Indistinguishable fermions cannot occupy the same total state. They stay away from each other. (All of this is observed to be so.) The electron is a fermion (actually, a fundamental fermion).

Recall from Chapter 12 Dirac's calculation that the electron has an intrinsic angular momentum, called spin, with values of either +½ or −½ times one basic unit of angular momentum. Electrons have *half* integral spin! Note also that the wavefunctions describing the spatial states of the electron are fuzzy and spread out. Electrons are *intrinsically* indistinguishable because they are the same kind of particle. And so, no two electrons can be in the same total state, that is, be in a place where their wavefunctions substantially overlap (in our visualization, but also mathematically), have the same energy, have the same spatial-state angular momentum, have the same magnetic component of the angular momentum, and have the *z* component of their spins in the same direction (i.e., both +½ or both −½) at the same time. If all of their other properties are the same, electrons can't be in the same place, and so they stay away from each other. This is exclusion.

We will now see how very important exclusion is in explaining the structure and properties of atoms having more than one electron.

HOW EXCLUSION WORKS IN EACH MANY-ELECTRON ATOM

Each atom, like hydrogen, has its own set of an infinite number of quantized states that its electrons may occupy. (The properties of 128 of these states have been represented in a generic way in Table II.) Complicated as each electron spatial state may seem from the appearance of its electron cloud, by the very nature of its being a solution of the Schrödinger equation, even in a multielectron atom, each state (mathematically) overlaps no other spatial state. And so electrons in the atom of an element can stay away from each other by being in separate spatial states.

The number of electrons in the atom of each element is by definition the atomic number of that element. So, for example, hydrogen has one electron, helium two, oxygen eight, neon ten, argon eighteen, and so on. And each electron, like most things in nature, tends to occupy the lowest energy state that it can in the atom of its particular element. However, because of exclusion, no two electrons can occupy the same total spin and spatial state in the same atom. So, the electrons tend to occupy, successively, the lowest energy state of each atom, and then the next lowest, and so on, one electron per state. The one electron in hydrogen tends to occupy its lowest energy state; the two electrons in helium tend to occupy the lowest and then the second lowest of the energy states for helium, and so on. Importantly, the chemical behavior of each element is determined largely (as we will see) by the last state to be occupied in its atom, and by the next lowest energy state that may be available for occupancy in that atom.

POPULATING THE STATES OF THE MANY-ELECTRON ATOM

Now we can see what states are occupied by the electrons in the atoms of the various elements (and eventually what that means for the properties of those elements). We proceed as follows, with reference to Table II.

There is some detail in the following numbered paragraphs. But in these specific examples we begin to show how quantum physics manifests itself in the properties of the elements and in the emergence of the periodic table. So please read with patience. The importance of this information will become clear later on.

(1) An electron tends to be in the lowest energy state available to it. Thus the overall *ground state of the atom* of each element results from the occupation by its electrons of the states of lowest energy for that atom, to the extent that those occupations are allowed by exclusion. An atom is usually found in this ground state.

(2) The single electron in a *hydrogen* atom, atomic number Z = 1, occupies either the [1s, +spin] or the [1s, −spin] state in the table for hydrogen; let's say the [1s, +spin] state located at the bottom left of Table I or Table II.

(3) For atoms with additional electrons (*because of exclusion*) no two electrons can occupy the same total (spatial and spin) state.

(4) So the first of the two electrons in a *helium* atom, atomic number Z = 2, occupies the lowest energy, [1s, +spin], state and then the second electron occupies the next lowest energy [1s, −spin] state in the helium atom's table of states represented by the generic Table II. This completes the occupation of the lowest (most-negative) energy states in the first (bottom) row of the table; what is called the 1s *first "shell"* of helium states for the helium atom. And helium is consequently *inert* and interacts with no other elements, as will be explained later in this chapter.

(5) The three electrons in a *lithium* atom, atomic number Z = 3, occupy its [1s, +spin] state, its [1s, −spin] state, and then its next lowest energy [2s,

+spin] state in *its* table of states for lithium (again, represented by generic Table II). And the four electrons in a *beryllium* atom, atomic number Z = 4, occupy these three of its electron states plus the [2s, –spin] beryllium state in its table of states (again, represented by generic Table II), completely filling the 1s shell and the 2s *"subshell"* part of the second row of states). (Note that a subshell is filled whenever all of the states within a given row of either an s, p, d or f block are filled.)

(6) Continuing with the five electrons of the Z = 5 *boron* atom, we see that its states are occupied up to and including the first of its plus-spin 2p states in the second row of a table for the boron atom.

(7) For the Z = 6 carbon atom and the Z = 7 *nitrogen* atom additional plus-spin, 2p states are occupied in their respective tables of states, each represented by our generic table of states, Table II.

Note that in each block of states at each energy level, all states of plus spin tend to be occupied before the beginning of the occupation of states having minus spin. That is because as soon as we begin occupying both plus-spin states and minus-spin states having the same *n*, *l*, and *m* quantum numbers, we have two electrons spatially virtually on top of each other. Because these electrons each have a charge of –*e*, they tend to repel each other but are otherwise trapped in their identical spatial-state clouds. This repulsion causes an increase in energy, and nature likes energy to be low and avoids such an increase as long as possible. So the electrons seek, subject to exclusion, to first populate within each block only a sub-block of either plus or minus spin. Consistent with our understanding that the lower energy states in any given row occur toward the left of the row, we have the plus-spin states populated first.

(8) And so it is that *oxygen*, for example, with atomic number Z = 8 and the first occupancy of a negative-spin 2p state (because it experiences the first population of a p block spatial state with electrons of both plus and minus spin), has some peculiar chemical properties relative to what might otherwise be expected (as will be discussed in Appendix D and Chapter 15).

(9) Skipping ahead, we find that the states of the Z = 10 *neon* atom are occupied by its ten electrons up to and through the entire 2p subshell, that is, through the entire *second*-row shell of states for the neon

atom. (And that is why neon is *inert*, as will also be discussed later in this chapter.)

(10) We continue with the filling of states for the atoms of the rest of the elements. In some cases the subshells and shells for those atoms are "just" completely filled with electrons. Those elements whose atoms have a just-filled subshell or shell have special properties (as we have already noted for helium and neon), as do those whose atoms begin to fill otherwise-identical states with opposite spin (as we have already noted for oxygen).

MARKING THE LAST OCCUPIED STATE TO CREATE A PERIODIC TABLE

For each element, we can mark the position of the state that is last filled with an electron by inserting the atomic number of that element into the square of Table II for that last-filled state. The result, if we do this for all of the elements, is Table III (which otherwise resembles Table II.) It would remain then only to: (a) add the corresponding name and symbol for each element to each square of Table III and (b) move the second square from the left in the bottom row (for helium, Z = 2) to the far right to create our periodic table, Table IV (shown also as Table B.2 of Appendix B). The move of the square for helium is shown by the arrow at the bottom of Table IV. (The reason for this move will become clear shortly.)

THE ELECTRONIC STRUCTURE AND CHEMISTRY OF THE ATOMS OF THE ELEMENTS

Table IV, our periodic table, was created as described in Appendix B initially as an arrangement of the elements sequentially in rows according to their atomic numbers and in columns by their empirically determined chemical properties. Table III, by contrast, has been created by marking the final occupancy of states of higher and higher energy levels (across sequentially higher rows) across blocks of states of particular angular momentum, sub-blocks of particular spin, and columns of the component of angular momentum that would align parallel to a magnetic field. *The similarity of these tables reflects the linkage between the observed chemical properties of the elements* (described in Table IV) *and the theoretically derived occupied electron-state structure of their atoms* (described in Table III).

Table III, of course, was put together through a semiquantitative understanding of the quantum mechanics of the many-electron atom. But the prediction from Table III of which type of state (s, p, d, or f), and how many states of each type, are occupied in the atom of each element is in good agreement with the predictions made in more sophisticated ways.[1]

Experiment has shown that our predictions are *100 percent accurate* for 66 of the 75 elements in the lower three blocks of the tables! And the predictions are *close to being accurate* for the remaining 9 of these elements, all of which lie in the central block containing the (so-called) transition metals, in which some of the d states are populated before both s states are populated. The situation is more complicated for the elements in the upper f block of the tables, where the energies of the last-filled of the various types of states are not as distinctly separated and some of the 4f and 5f states are populated before all of the 4d and 5d states are populated. In spite of these differences, overall, our crude method of projecting the electronic structure of the elements from what we know of the hydrogen atom has been remarkably successful.

THE NOBLE GAS ELEMENTS

We now examine particularly the electron structure of the *inert elements, the noble gases* (those elements in the last column in both tables). The noble gases are among those elements for which our projection of the occupation of states has been shown by experiment to be 100 percent accurate.

We know that the atom of each element (labeled in Table III only by its atomic number) consists of electrons that occupy, one per hydrogen-like state, all of the states shown in all of the squares up to that point in its equivalent of Table II. Now (remembering that we represent the atomic number of each element generally by the letter Z), we note particularly that the square second from the left at the bottom of Table III marks the filling of both of the 1s states in the atom of the element helium, Z = 2.

As will be discussed more later, this filling of both states in the lowest-energy-level first row, results in a highly compact atom with its electrons so tightly bound that they cannot be pulled loose by other atoms. Nor does the helium atom seek to gain an electron by combining with the atoms of other elements. That is because the two positive charges of the protons in the helium atom's nucleus are almost entirely shielded by the negative charges of its two electrons, so that the nearest unoccupied state that the new electron might go into is only very slightly negative in energy and doesn't attract and hold an electron. Helium, thus, as observed, tends to be inert and to combine with no other elements. So in helium we have the complete filling by its two electrons of what we call the first *shell* of states.

The atoms of all of the elements down the rightmost column of Tables III and IV, with atomic numbers Z = 10, 18, 36, 54, and 86, all have completely filled shells in which the electrons are similarly expected to be relatively tightly bound (as will be discussed further in Chapter 14). These shells also almost completely shield the positive charge of the protons in the nucleus so that they do not attract additional electrons. These, the rest of the noble gases, might also therefore be expected to be inert. But, as one considers the heavier of these noble gases, those elements of higher atomic number, their outer electrons, though completing a shell, are less and less tightly bound. These outer electrons can be stripped away by such highly acquisitive elements as fluorine and oxygen. So these heavier noble gases are not inert. Regardless, from the filling of states in Table III (subject to further discussion in Chapter 14), we have answered the questions posed earlier as to why elements with atomic numbers Z = 2, 10, 18, 36, 54, and 86 tend to be inert, and why we have the curious spacings of the number of elements before or between the listings of these noble gases in Table IV, that is, we have answered why the numbers of elements in the first six rows of the periodic table are 2, 8, 8, 18, 18, and 32.

Now we can also understand why I've moved the square for helium, Z = 2, from the left to the right side of Table IV, as shown by the arrow at the bottom of the table. The helium atom has a completely filled shell of states, just as do the atoms of the rest of the *noble gases* listed in the last column of this table. (But to keep the electron structure straight, we must remember that both electrons in helium are in an s state, despite helium's listing in the otherwise last-filled p state Group VIIIA column of Table IV.)

ELECTRON STRUCTURE AND CHEMICAL BEHAVIOR OF ATOMS CLASSIFIED IN "GROUPS"

Now we begin to see how the occupation of states by electrons in atoms explains the chemical properties of the elements. To start with, note that the total number of electrons acquired or shared by one or several atoms when they react (combine) to form a *compound* is always equal to the number of electrons that are taken away from or shared by one or several others.

So, for example, for the compound water, H_2O, two hydrogen atoms each give up or share one electron to combine with one oxygen atom, which acquires or shares two electrons to completely fill its p subshell and its $n = 2$ shell of states. The organized listing of elements whose atoms tend to acquire or give up electrons is inherent in the layout of Table IV and is called out in the labeling of its columns.

The strength toward acquisition is highest for atoms of those elements shown in the bottom rows and (except for the last column) to the right of the tables, in those squares that are *darkly shaded*. It is the nearness to a completely filled (lowest-energy-configuration) shell that makes these elements particularly active in acquiring an electron or electrons. These are the *nonmetals*. (As an example, note that atoms of fluorine [F, Z = 9, at the bottom of the Group VIIA, second from the right column] are so reactive that they displace oxygen from water to form hydrofluoric acid, HF, which is so reactive that the fluorine in it will further rip silicon [Si, Z = 14, second from the bottom in the much less acquisitive Group IVA column] from oxygen in SiO_2 to etch glass.

Conversely, atoms having *one more* electron than fills the row of a p block (called a valence subshell) want to be in a just-filled valence subshell configuration and tend to react strongly to *lose* that last electron. That is because the electron outside of the filled subshell is largely shielded from the attraction of the nucleus by the smaller, essentially spherical, ball of negative charge of the filled subshells within. The elements having these kinds of atoms are marked in the first column of Table III and are listed in the first, *Group IA*, column of the periodic table. The elements whose atoms have *two extra* electrons beyond the filled subshells tend to less aggressively try to lose these electrons and are marked in the second column of Table III and are listed in the *Group IIA* second column of the periodic table. The tendency of atoms of all of the elements to shed electrons increases with their location farther up and to the left in the tables.[2] Those elements that tend to lose electrons are shown *without shading* in Table IV. They are classified as *metals* because they show this tendency. They have what we know as metallic properties: electrical conduction, luster, and (usually) malleability.

(As one familiar example of a reactive metal, consider sodium [Na, Z = 11, third from the bottom in the leftmost Group IA]. We know sodium from its combination with chlorine to form sodium chloride, NaCl, common table salt. Sodium is a malleable metal, an electrical conductor with metallic luster. But thrown into water, sodium will rip one oxygen along with one hydrogen atom from the other hydrogen atom to produce sodium hydroxide, NaOH. If enough sodium is present, this will create a violent reaction with heating and steam, releasing hydrogen to possibly ignite and burn again or even explode, should it accumulate in sufficient concentration and come into contact with the oxygen in the air. Potassium [K, Z = 19] will displace the sodium from NaOH, and so on. Going the other way, down the first column, lithium and hydrogen are not so reactive; and beryllium, at the bottom of the Group IIA, alkaline earth metals column, is only reluctantly metallic. We don't think of hydrogen as a metal because we see it mainly as a gaseous element that doesn't display metallic properties; however, theoretically, under enough pressure and at low-enough temperatures, hydrogen may be solid and display the properties typical of a metal.)

The atoms of a few of the elements in the first four columns in the p block of Table III and shown *lightly shaded* in a diagonal from bottom left to top right across the Group IIIA, IVA, VA, and VIA columns of the periodic table, have properties that lie "in between" those of the metals and those of the nonmetals, as will be discussed in later chapters. These are known as *metalloids* or *semiconductors*. (We are familiar with silicon and germanium and their use in modern electronics.)

The letter *B* after the Roman numerals labeling the columns across the left–center of the periodic table refers to the additional groups (columns) that occur when electrons occupy d states after s states outside of a completed p block subshell. The elements in these groups are the *transition metals*. (Among these are iron, nickel, and chromium, which are used to make steels, and some of the elements sometimes found in their uncombined metallic state and known since ancient times: copper, silver, and gold.)

The Row #6 f block of lanthanide elements shown near the top of the table, also known as the *rare earths*, is to be inserted into the Row #6 transition metal series, after lanthanum, Z = 57. The Row #7 f block of actinide series elements shown at the top of the table, also known as *heaviest elements*, is to be inserted into the Row #7 transition-metal series, after actinium, Z = 89. (Among these are radioactive uranium and plutonium, familiar to us through their use in the production of nuclear power and weaponry.)

As discussed earlier, the s and d (and in Rows #6 and #7 also f) electron states in some of the elements across the B-type groups (columns) in each row may have nearly comparable energies, so that the order of their occupancy is not as clearly given as for elements in the A-type groups. And, despite their similar Roman numerals, the electronic makeup and properties of the atoms of the elements in the similarly numbered A-type and B-type groups may sometimes be quite different in their properties. The B groups contain three more columns of elements than do the A groups, accommodated by having all three of the columns in the center of Table IV labeled as VIIIB. None of the elements in these Group VIIIB columns bear any resemblance to the noble gases of Group VIIIA. Note particularly that the lightest transition elements in these three columns of Group VIIIB, those in Row #4, are iron, cobalt, and nickel, which all tend to be magnetic.

(More is described of the transition elements in Appendix D. However, Appendix D is best read in light of the physical picture of what goes on in the many-electron atom, as presented next.)

Chapter 14

THE PHYSICS UNDERLYING THE CHEMISTRY OF THE ELEMENTS

Remember, all of our considerable understanding thus far of the electronic structure of the atom is based on our crude construction of tables from aspects of quantum mechanics: (a) the raw spatial-state solutions of Schrödinger's equation for the hydrogen atom (some represented by the "clouds" shown in Fig. 3.8 and Table I); (b) the +½ and –½ electron spin states found by Dirac; (c) exclusion; and (d) an additional consideration for the screening of an electron from the charge of the nucleus by the charges of the other electrons in the many-electron atom.

In this chapter we see how quantum mechanics goes much further than these tables in explaining the fascinating diversity of chemical and physical properties of the elements and their tendency to bond.

I sincerely expect that most of you will emerge from the reading of this chapter with a sense of wonder at the awesome beauty of nature. But I do need to let you know that we are about to discuss the physics of the many-electron atom in part on a conceptual level beyond what is taught in most introductory physics and chemistry courses.

From what you have already learned, you have an excellent background for understanding this material. However, should this chapter turn out to be a bit too "heavy" or detailed for you, I would suggest that you skim a bit to get a sense of what is presented here. If you decide to skim through, I suggest the following: read the unindented paragraphs, but skip Figure 14.1 and the indented paragraphs, and then read the sections titled "Putting It All Together," and "Introduction to Appendix D" in the concluding paragraphs of this chapter, before you move ahead to Chapter 15.

THE HYDROGENLIKE PHYSICS OF THE MANY-ELECTRON ATOM

Here we deal conceptually with what happens to the states of the electron in the atom as we consider atoms of successively higher and higher atomic number, that is, atoms having more and more electrons (and, of course, by definition of *atom*, always in each case an equal number of more and more protons). Then we use this conceptual framework to understand the measured physical sizes and the chemical properties of the elements. I mainly describe groups of elements, but I highlight here and there the physics and chemistries of a few of the more interesting elements.

Recall that the electrons in an atom of any element have their own set of possible states somewhat similar to those described for hydrogen in Table I, each set with its own relative sizes somewhat similar to those shown for a few of the states of hydrogen in Figure 3.8. In principle (and, for the atoms of some of the elements, in practice) these states are the solutions (with the aid of computers and numerical approximation) of Schrödinger's equation for the many-electron atom. For these solutions, in addition to taking into account the *electrostatic attraction* of the negatively charged electrons to the positively charged protons in the nucleus, the equation includes the *electrostatic screening* (shielding) of this attraction from each electron by the negative charges of the other electrons (to the extent that the other electrons lie between that electron and the nucleus). The effective nuclear charge that the electron then sees is the charge of the protons minus that part of the charge (on average) of all of the other electrons that (for the spatial states that they occupy) lie between our electron and the nucleus.

Remember that in Chapter 13 we developed a generic table of states, Table II, for the many-electron atom, by recognizing in a qualitative way that the degree of penetration of each electron through this screening or shielding (i.e., the fraction of the nuclear charge that our electron still sees despite the screening) is different and shifts the energy levels differently depending on whether the electron is in an s, p, d, or f type of state. In the indented paragraphs below we examine in further detail how this happens, considering the physics of electron states and their energies. I generalize the results in unindented paragraphs that follow afterward.

(A) As we consider first the hydrogen atom and then successively the atoms of higher atomic number, we find that the *states* for each successively heavier atom are drawn down (compared to those of the preceding atom) to successively lower energies and smaller sizes by the increased attraction of the additional proton in each successive atom's nucleus. To begin to quantify this, we look first at the attraction of just one electron to a nucleus of Z number of protons, without the complicated screening that would result from the presence of the other electrons. (We will bring the effect of these other electrons into the picture later on.) Schrödinger's equation for the energies of the spatial states of the single electron can be solved exactly and gives us the formula

$$E = (-13.60\ \text{eV}) \times (Z)^2/(n)^2,$$

where the multiplier $(Z)^2$ shows the greater negative binding energy (with an associated greater attraction) of the larger number of protons. (This is the same expression that we used in Chapter 11 to calculate the energy levels for the $Z = 1$ hydrogen atom, but there "1" was substituted for Z so that we didn't see Z in the formula. Remember, just as described for the hydrogen atom in Chapter 11, there are $2 \times n^2$ combined spatial and spin states at each energy level n.)

In Figure 14.1, I show the results of calculations using this formula for three simple cases, each involving only one electron.

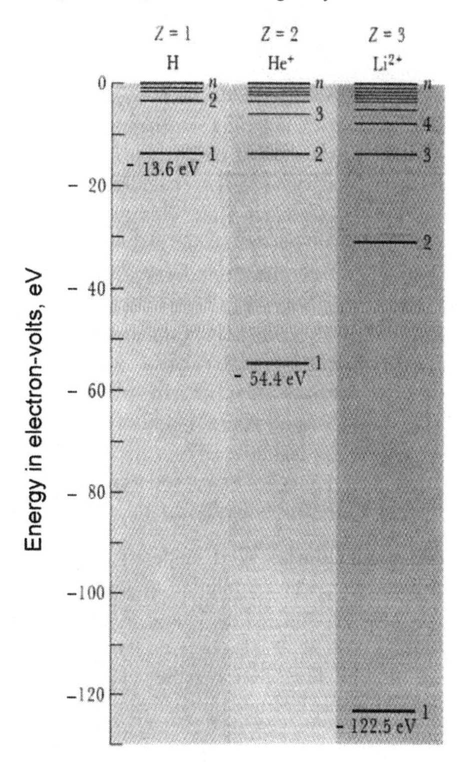

Fig. 14.1 Energy levels for single-electron atoms and ions: for the $Z = 1$ hydrogen atom (first column), the singly ionized $Z = 2$ He^+ ion (second column), and the doubly ionized $Z = 3$ Li^{2+} ion (third column). Note the four- and nine- times lower energies at all $n = 1, 2, 3$, and so on levels for the two ions, respectively. (Figure 14-11 from Fine and Beall, Reference E, with permission from Dr. Leonard W. Fine.)

Case (1) is the Z = *1 hydrogen atom*, which of course is one electron around a nucleus of one proton. The negative energy of the *n* = 1, 1s, lowest energy, most tightly bound state of the one electron is (plugging in 1 for Z and 1 for *n* in the above formula):

$$E = (-13.60 \text{ eV}) \times (1)^2/(1)^2 = -13.6 \text{ eV},$$

as shown by the lowest horizontal line in the first column of the figure.

The energies for the first few of the higher-*n*, higher-energy states for the electron in hydrogen are also shown in this first column. (We also calculated and tabulated these energy levels earlier, in Chapter 11.) Only the *n* = 1, *n* = 2, and *n* = nth of these levels are labeled in Figure 14.1. We show only a few of the infinity of states because the higher-*n* states become bunched closer and closer together as they approach a zero-energy level, so there is no room to include or label them. The labeling of the nth, topmost, state shown is meant to indicate that there are more and more states, an infinite number, with states eventually getting infinitesimally close to the zero level of energy as *n* approaches infinity. At zero energy, the electron is free, unbound, and no longer attached to the atom. As *n* approaches infinity and the energy levels of the states approach zero, the electron in these states is more and more weakly bound, and the probability clouds for these states are larger and larger in size (until the probability cloud for the electron is spread essentially everywhere, and the electron can be almost anywhere, is essentially no longer bound, and breaks free of the atom).

Case (2) is the Z = *2 singly-ionized* He^+ *helium ion*, that is, the helium atom with one of its electrons removed so that the resulting ion consists of one electron around a nucleus having two protons. (Note that we continue to label an ion with the Z atomic number of the atom from which it is formed. Note also that the unbalanced plus charge [the *valence*] of the ion is indicated in the superscript following the symbol for the element. In general, the valence indicates the charge imbalance of an ion, plus by the number of electrons shy of matching the number of protons in the nucleus of the ion, minus by the number of electrons exceeding the number of protons in the nucleus.) The energies that are calculated for the electron states for the helium ion are shown in the second (lightly shaded) column of Figure 14.1. Note that the *n* = 1, 1s, lowest-energy state is

$$E = (-13.60 \text{ eV}) \times (2)^2/(1)^2 = -54.4 \text{ eV},$$

four times the negative energy of the comparable state for the electron in the hydrogen atom.

Case (3) is the *Z* = *3 doubly-ionized* Li^{2+} *lithium ion*, with two of its electrons removed so that the resulting ion has one electron around a nucleus having three protons. The energies that are calculated for the states available to this electron are shown in the third column of the figure. Note that in this case the energy of the *n* = 1, 1s, lowest-energy state is at

$$E = (-13.60 \text{ eV}) \times (3)^2/(1)^2 = -122.4 \text{ eV},$$

nine times the negative energy of the comparable state for the electron in the hydrogen atom.

(B) Comparing the energy levels for each quantum number *n*, we see in Figure 14.1 that the energy in each successive column is lowered further in proportion to the number of protons in the nucleus squared.

This is most easily seen by comparing the $n = 1$ energy levels for each of the three cases. A correspondingly greater electrostatic attraction of the electron to the nucleus and consequent tighter and tighter binding of the electron (smaller hydrogenlike probability-cloud size) occurs in each successive case. And this continues for heavier and heavier elements. Electron transitions from higher energy levels to the extremely low energy levels for the inner-electron states in the atoms of copper, $Z = 29$, and tungsten, $Z = 74$, is demonstrated dramatically in the production of x-rays (as described in Appendix E).

(C) When the removed electrons are replaced in the helium and lithium ions, so that we address them now as atoms with two and three electrons, respectively, the energy levels shown in the second and third columns of Figure 14.1 will be shifted upward to reflect the effectively smaller charge of the nucleus that is caused by the mutual screening of these added electrons. (By *screening* we mean that on average the other electrons spend some of their time partly between our electron [at whatever location in its probability cloud it might be on average] and the nucleus, so that our electron sees a net smaller positive charge equal to that of the positively charged protons minus that of the on average interfering fraction of the negative charges of the other electrons.)

(D) Similar proton attraction and screening effects occur for the electrons of the many-electron atoms of higher atomic number, that is, in atoms of the rest of the elements.

PUTTING IT ALL TOGETHER

And so we have an *overall picture* of what happens as we consider atoms of higher and higher atomic number. Nature always tends to seek an overall lowest-energy situation. *For each successively higher-Z atom, electrons occupy successively one more state* starting from the lowest and extending to higher and higher (less-negative) energy levels and larger and larger corresponding sizes *within each atom's ladder of energy levels. At the same time, the whole ladder of energy levels is drawn down* to lower energies and smaller sizes by each added proton in the nucleus. It's not an even process because the screening of the nucleus is uneven as the s, p, d, or f subshells of states are filled. It's as if the additional electron in the atom of each successive element reaches the next unevenly spaced (energy) rung on a ladder, while at the same time the entire ladder of state energy levels is being drawn downward and somewhat stretched because of the additional screened charge of the added proton in its nucleus. So, the topmost electron for one element on *its* ladder isn't a whole lot higher in energy and larger in size (if higher in energy and larger at all) than the topmost electron from the preceding element on *its* ladder. This is true even though the atom of the succeeding element has one more electron and this electron is on a higher-numbered rung of its ladder of energy levels.

As you can see, it's complicated. No wonder it has been so difficult to solve Schrödinger's equation for atoms having more than a few electrons, even by approximation methods using powerful modern computers.

INTRODUCTION TO APPENDIX D

Fortunately, as I describe in Appendix D, there is a way of directly measuring the energy levels of the outermost occupied states of the atoms of all of the elements. These measurements explain much of the chemical properties of the elements, the sizes of their atoms and the arrangement of the periodic table. (And with these we also see how the measured sizes of atoms are actually smaller and smaller as we successively consider heavier and heavier elements with more and more electrons within each period of the periodic table. In our periodic tables, each period is the span of a single row of elements.) Appendix D is a fascinating, partly visual, and triumphant conclusion to all that we have learned so far, but it's a bit too detailed to be presented here. Even though I have removed it from the main flow of the book, I nevertheless strongly encourage you to at least scan through to see what is there.

And realize that (from the quantum framework of the present chapter and earlier parts of this book) you already have the deep understanding of why we get these measurements and associated chemical behaviors. Specifically, our reading until now has provided an exceptional view into what electrons are; how the electron states get their sizes, their shapes, and these measured outer energies; and why the energies and associated chemical properties vary periodically as we consider successively atoms of higher and higher atomic number.

We see next how the properties of the outer electrons of the atoms of the elements determine some of the chemical bonds that may form between atoms, using just a few familiar, interesting examples of the metals and compounds that result from such bonding.

Chapter 15

A FEW TYPES OF CHEMICAL BONDS, FOR EXAMPLE

> ***"The explanation of how atoms combine to form molecules and why the resulting molecules have the unique shapes that identify them is one of the major triumphs of quantum mechanics."***
>
> *—Fine and Beall, Chemistry for Engineers and Scientists (Reference E, p. 544)*

With bonding we enter the realm of chemistry, the examination of how it is that elements combine. We are concerned about this phenomenon here because the types of bonds that may be formed are related to the basic electronic structure of atoms involved as explained by quantum mechanics.[1] Bond shapes, molecular shapes, the polarization of molecules and intermolecular bonds, and the properties of the liquids and the crystal structures, and the mechanical properties of the solids that may be formed from atoms are all influenced, if not determined, by their electronic structures. And so I want to provide just a little bit of an illustration of bonding, simplified, just to give you a taste of what might lie ahead in chemistry.

I first present a very simple description of ionic, covalent, and metallic bonds, essentially definitions. (To better understand these bonds, you may wish to read the beginning of Appendix D through the second section, "The Propensity to Bond.") I then go a bit further to describe a few examples. To do this I consider the atoms of mainly three elements: oxygen or carbon atoms in molecular combination with atoms of hydrogen.

BOND TYPES

An *ionic* bond forms when an atom loses one or more electrons to become a positively charged *cation* while another atom acquires the electrons from the first atom to become a negatively charged *anion*, and the cation and anion are then electrostatically attracted to each other to bond together to make a molecule. One example is sodium combining with chlorine to form sodium chloride, NaCl, common table salt.

As described in Chapter 14, what drives the bonding is the reduction in overall energy when the atoms acquire or lose electrons so that low-energy just-filled shells of electrons are produced in the resulting ions. (And, remember, things in nature tend to be in states of lowest energy.) Several cations and anions can be combined together, as long as the number of electrons lost by the atoms to form cations is the same as the number of electrons acquired by atoms to form anions. For example, two atoms of aluminum may each lose three electrons, forming cations in reaction with three oxygen atoms, each of which acquires two of the electrons, forming the compound aluminum oxide, Al_2O_3. (With trace amounts of different transition-metal impurities replacing some of the aluminum atoms, the compound may take on a blue color, forming a sapphire, or a red color, forming a ruby.)

Some atoms like to share electrons with other atoms to form *covalent* bonds, as occurs for example when two hydrogen atoms bond to form a hydrogen molecule, the molecule making up hydrogen gas. In this case, for example, each hydrogen atom "thinks of itself" as having two electrons to complete a heliumlike completed first electron shell.

All bonds are some combination of ionic and covalent. But there is much, much more to it and this is an oversimplification. For example, there is the *metallic* bond. The billions of atoms in a metallic solid (or liquid, as for example with mercury) are held together by a "sea" of outer electrons from all of the atoms in the liquid or solid. Although the number of electrons locally balances the positive charges of the protons in the nuclei of each of the atoms, the outer electrons of each atom are essentially free to move around within the solid or liquid, and it is these electrons that can carry electrical current.

And there are other bonding mechanisms that I haven't described. All of these many mechanisms can be spread with varying amounts of applicability across the whole spectrum of the electronic structures of the elements. This is the large palette that nature and the chemists have to work with in creating the multitude of compounds and substances that surround us. (And most of what surrounds us are compounds.) We present briefly now just a few examples of the bonding of atoms and the chemistry of molecules.

CARBON, THE CHAMELEON!

Carbon is perhaps the most versatile of elements. Shown darkly shaded with the nonmetals at the bottom of Group IVA in the periodic table (Table IV), carbon is nevertheless a border element, bonding with either metals or nonmetals.

The carbon atom consists of *two* 1s electrons, *two* 2s electrons, and *two* 2p electrons surrounding a nucleus of six protons and six neutrons, although some isotopes have more neutrons. The 2s and 2p states are so close in energy that they can collectively distort to form *hybrid*[2] states, to accommodate particular bonding situations. And a carbon atom bonds readily with another carbon atom. Partly for this reason, carbon forms more compounds than any other element. It is the core element of organic chemistry and the key element of biology and all of life.

Carbon, in its most abundant naturally occurring form (*allotrope*), is graphite or graphitelike, where each carbon atom is bonded to three others in sheets of connected hexagons, with carbon atoms at each "corner" of the hexagons. Isolated sheets are the allotrope of carbon known as graphene, which is described along with related "fullerenes," buckyballs, and nanotubes in Chapter 22. The tight bonding of carbon atom to carbon atom within the sheets results because one 2s and two 2p electrons *hybridize* to provide connections equally to the three neighboring carbon atoms. Graphite is thus soft and slippery because the sheets can slide over each other. Meanwhile, each carbon atom's fourth $n = 2$ level electron is free to wander around to carry an electrical current when driven to do so by a battery or other voltage source. So, graphite is electrically conducting.

In contrast, at high temperature and pressure, the carbon atom's two 2s and two 2p electrons hybridize into a lobed tetrahedral arrangement that allows each carbon atom to bond with four others to make diamond, the hardest substance known to us, and also a perfect electrical insulator (there are no free conduction electrons). This same tetrahedral bonding is evident in the compound methane, CH_4, the major constituent of natural gas, where each of four hydrogen atoms bonds by sharing its electron into one of the four equally spaced tetrahedral lobes of the hybridized carbon states.

CROOKED CONNECTIONS

While the H_2O molecule in liquid water or ice tends to be drawn into a tetragonal arrangement with the help of noncovalent intermolecular hydrogen bonds with its neighbors, the isolated H_2O molecule is bent to nearly a right angle: that is, the two hydrogen atoms attach to the oxygen atom at a relative angle of 104.5 degrees, as shown in Figure 15.1(c). (Many other molecules show a similar behavior of connections at odd angles.)

This bent structure was first explained over seventy years ago by Nobel Prize–winning chemist Linus Pauling in his work on the nature of the chemical bond.[3] The perpendicular orientation of dumbbell-shaped oxygen 2p spatial states (such as the 2p spatial state for the hydrogen atom shown at the bottom of the second "column" of Figure 3.8) is the root cause for this bent shape. Figure 15.1 shows the oxygen atom, starting at the bottom left, as a more-simply-drawn, "unfuzzy" depiction of two oxygen 2p states, one with its lobes along the *x* axis and one with its lobes along the vertical *y* axis. The (a), (b), and (c) steps described in the caption show the formation of the bent molecule. (More recent calculations show that this picture of the bonding isn't quite correct,[4] as the oxygen states involved in the bonding are somewhat *hybridized*[5] with the 2s state, but the influence of the 2p states in providing a directionality for the bonds is still there.)

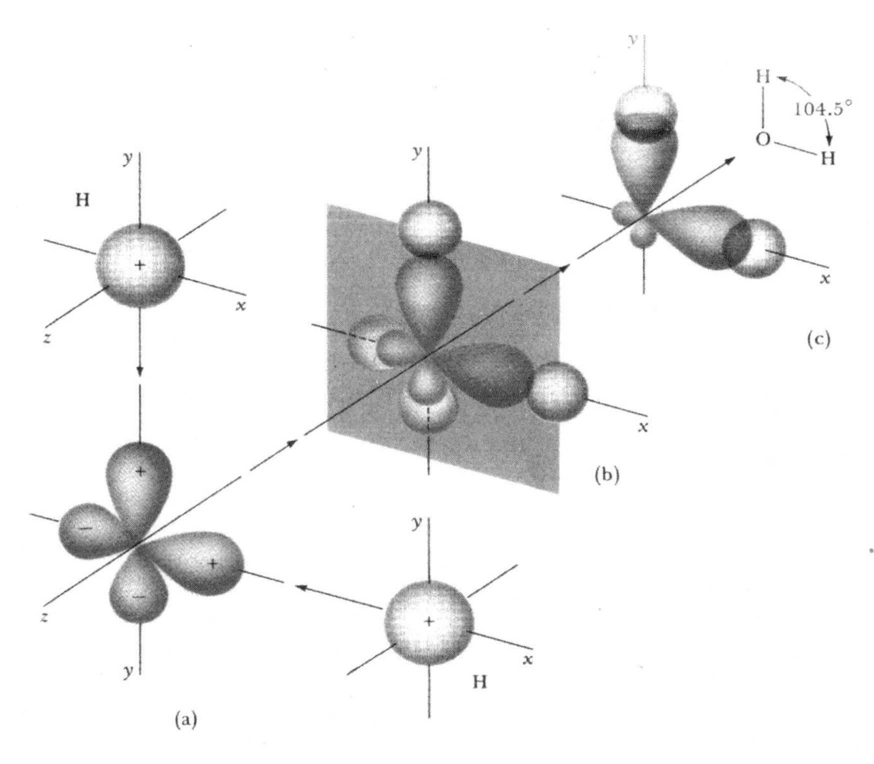

Fig. 15.1. The isolated H_2O water molecule's bent shape. (a) The hydrogen atoms with their 1s spherically symmetric spatial states approach the oxygen atom with its two perpendicular dumbbell-shaped 2p spatial states along the *x* and *y* axes. In (b) the states draw together and in (c) their spatial states begin to overlap. Also shown in (c): the mutual repulsion of the clouds of electrons that are drawn toward the hydrogen atoms spreads the angle from 90 degrees to 104.5 degrees. (Figure 18-2 from Fine and Beall, Reference E, with permission from Dr. Leonard W. Fine.)

Odd angles in the structure of the molecules of many compounds are explained by the strange shapes of the various atoms' spatial states that result from the quantum mechanical solutions of Schrödinger's equation. There is simply no classical explanation for these angles.

Chapter 16

THE MAKEUP OF SOLID MATERIALS

Here we provide some insight into the fascinating and varied structures of solids as background for the discussions of the many applications in the chapters of Part Five. We start with the *atoms* of the *elements* as the basic building blocks of all substances. Atoms of some elements will chemically combine (as we've discussed) in definite ratios to form the *molecules* of *compounds*, like aluminum and oxygen combining to form aluminum oxide, Al_2O_3, where two aluminum atoms chemically bond together with three oxygen atoms.

Most solids are *crystalline*. They have atoms or molecules that stack themselves in neat three-dimensional lattices called *crystals*, like a box of carefully, tightly packed oranges all of the same size. Quartz, which may have the same chemical composition as glass, is crystalline. Diamonds are crystals of carbon atoms. Sapphires and rubies are crystals of Al_2O_3 molecules, crystals that take on different colors depending on the presence of a relatively small number of atoms of other elements as *impurities*. *Gems* are cut or cleaved from large single crystals, which have one stacking order and one stacking orientation throughout.

Often substances are *polycrystalline*, consisting of millions of small crystalline grains that have grown together as a molten material solidified, with each grain having the same internal stacking of atoms, but with each grain having a different orientation.

In some cases the atoms are bonded in some way to their nearest neighbors, often forming hard crystals that tend to be electrical insulators. *Metals* (composed of just one metallic element) are held together in a *metallic bond* as solids or liquids, that is, by a "sea" of electrons spread out evenly but not attached to any particular atom, though attracted to the nuclei in their vicinity. Because in many solid metals the stacked layers of atoms can easily *shear* over each other (like the cards in deck of playing cards), these metals are malleable and ductile and can be rolled, drawn, or hammered into various shapes. When this is done, structural defects are created within the grains which prevent the shearing, and thereby *work-harden* the metal. (If you've ever bent a metal clothes hanger back and forth until it hardened, became brittle, and broke, you've had a demonstration of work-hardening.) The metal can then be *annealed* by heating it to a temperature where (without any melting) new grains form and grow to larger sizes, gathering atoms from the old grains at the boundaries between them and shifting the boundaries in the process. In this way, the older grains with the defects lose atoms to the growing new grains and eventually disappear, producing a relatively soft material once again.

The molecules in crystalline *compound* solids not only are neatly stacked but also tend to be bonded together molecule to molecule, producing hard, brittle substances.

Other solids are *amorphous*: the atoms or molecules comprising the solid have no particular stacking order. They are like very thick viscous liquids, so viscous that they tend to retain their shape. Glass is a good example. But, if we were to wait long enough, panes of glass would eventually, very slowly, flow like liquids down into a puddle.

Solids, liquids, and gases are different *phases* of a substance. The atoms or molecules of some solid substances can stack themselves in two or more different geometrical arrays. We say that these different stackings are different *solid phases* of that substance. Solids having different stacking orders are distinctly different phases of a substance, just as are melted *liquid phases* or evaporated *gas phases*. (I hope that you are unfazed by all of this!)

The atoms of more than one metallic element can often be combined into what is called an *alloy*. This is often accomplished by melting the elements together to form a *solution* (like sugar dissolved in water), which in the case of an alloy sometimes solidifies as a single-phase *solid solution* but often solidifies in such a way that grains of two or more different solid-solution phases are produced, each phase having its own separate proportions of the different elements and its own form of stacking. Sometimes, if the alloy is cooled fast enough from the melt or anneal, these second phases don't have time to form but can be caused to appear during later heat treating (heating but not melting) of the alloy. Bronze, for example, is an alloy of copper, tin, and zinc. Sometimes an alloy contains nonmetallic elements, and some of the phases can then also include a stacking of molecules (much like the stacking of molecules of the compound Al_2O_3 described earlier). Steels, for example, are alloys of iron and carbon and/or other elements (e.g., chromium and nickel for stainless steels).

For just a moment, I will discuss steel to illustrate some of *the sophistication of modern man-made materials*. Steels typically contain several different finely intermingled solid phases, each of a different chemical composition depending on the initial overall chemical composition of the steel alloy, the history of its melting, the rate of cooling to a solid, and then the *heat treating* to cause different phases to form. So it is that metallurgists have, for example, been able to create steels that have malleable phases that can be easily rolled into thin sheets and then cut into the form of razor blades, only then to provide a heat treatment that *precipitates out* finely divided solid phases within the original phases. These precipitates (like nails driven through a deck of cards to prevent the cards from sliding [shearing] over each other) make the overall alloy hard and stiff. For example, razor blades are made from steel that is soft when it is rolled into thin sheets and cut to make the blades, but then that same steel is hardened with a heat treatment that makes it easy to sharpen the blades by grinding and makes them slow to wear out and become dull.

Some phase changes can even take place suddenly at low temperatures without heating (by the shearing of one type of crystal lattice to form another). And so it was that some of our Liberty ships carrying cargo during WWII were found to break apart, because of unanticipated phase changes, when their hulls (of the wrong composition of steel) became exposed to the frigid cold of the North Atlantic and transformed into brittle materials.

In the next chapter we'll consider the electrical properties of solids. It is often both the electrical and mechanical properties that need to be considered in the invention of new materials and devices, as described in Part Five of this book, soon to follow.

Chapter 17

INSULATORS AND ELECTRICAL CONDUCTION IN NORMAL METALS AND SEMICONDUCTORS

Charles Kittel notes, in his book *Introduction to Solid State Physics*: "The difference between a good conductor and a good insulator is striking. The electrical *resistivity* of a pure metal may be as low as 10^{-10} ohm-cm. . . . The resistivity of a good insulator may be as high as 10^{22} ohm-cm. This range of 10^{32} may be the widest of any common physical property of solids."[1] This comment on such a remarkable property range applies to what we call *normal* materials. The range is expanded even further when we consider superconductors in Chapter 19, just ahead. But to appreciate superconductors, we first need to understand metals, insulators, and semiconductors.

METALS

The elements listed toward the left and top of Table IV in Chapter 13 or Table B.2 in Appendix B (those elements in the region without shading) are all classified as metals. Most of these elements in solid or liquid form (mercury, as an example of the latter) can carry an electrical current.

As also noted before, the atoms of metals in either solid or liquid form are held together by a collective *sea* of all of their outer electrons. *Thinking classically*, these electrons are essentially free to move around within the solid or liquid, and it is these electrons that can carry electrical current. They are spread more or less evenly throughout the metal, and their electrical charges attract them to locally balance the positive charges of the ions that they left behind when they broke loose from their atoms to roam free.

Because some of these metals are malleable and ductile, they can be pressed into shapes or drawn into wire. When used electrically, the wires are usually connected with other electrical components, which include some sort of power source that forces electrons to flow through a complete circuit. For example, in a flashlight they are repelled from the end of the negative terminal of a set of batteries to pass through a wire and a light bulb (circuit element) and back through another wire to the attracting positive terminal of the battery set.

Though it is handy to use this classical view descriptively, there is a problem with it: individual electrons, as particles, would be expected to collide with each other and the lattice array of ions, with the result that the metal would exhibit a much higher resistance to current flow than is actually found.

THE QUANTUM VIEW OF ELECTRICAL CONDUCTION: METALS, INSULATORS, AND SEMICONDUCTORS DEFINED

In the quantum view, all of the electrons are indistinguishable one from another and it is really impossible to know which electron is where, except that we do know that integral multiples of one electron charge, *e*, move out of the negative terminal of the battery, and the same amount of charge simultaneously moves into the positive terminal of the battery at the other end of the circuit. When Schrödinger's equation is solved for the entire collection of all of the electrons in all of the atoms, what results is a "band" of states at a very large number of very closely spaced energy levels.[2] Just as for the individual atom, the indistinguishability of electrons and their ½ spin requires that no two electrons may be in the same state. When all of the electrons in the metal settle into the lowest of the energy states, one electron per state, the band is filled to what has been called a *Fermi level*.[3] States having electrons moving in different directions are equally occupied, so there is no net flow of electrons.

When an electric potential, a voltage, is applied to the metal, some of the electrons are shifted preferentially to states that lie above the Fermi level, having motion in one particular direction, and so there is a net flow of electrons in that direction and the conduction of an electrical *current*.

But the lattice of ions causes gaps in the band of available states, *band gaps*, energy regions for which there are no solutions to Schrödinger's equation.[4] Electrons simply cannot exist at these energies.

If the electrons occupy the available states to a Fermi level that is comfortably below or above one of these band gaps, the shift in states for net motion can easily occur. The electrons near this Fermi level are said to be in a *conduction band*, and they are impervious to the lattice of ions. The electrons also tend not to interfere with each other, because they are all subject to exclusion and in separate states.[5] Conduction can easily take place, and one has a *metal*.

If the electrons occupy the available states to a Fermi level that just reaches one of these band gaps, and if the gap is sufficiently large (larger than the applied voltage, so that the voltage cannot lift them in energy to states above the gap), their part of the band is essentially capped, there are no states that the electrons can shift into, there is no electrical conduction, and one has an *insulator*. That part of the overall set of bands that they fill is called a *valence band*.

If the electrons occupy the available states to a Fermi level that is just below or above one of the band gaps, that is at the top of a valence band or at the bottom of a conduction band, or if the gap isn't too large so that the electrons can be excited thermally or by the application of voltage across the gap to the conduction band above, one has a limited amount of conduction and what is called a *semimetal* or *semiconductor*, with characteristics that will be described in Chapter 23.

DIRECT CURRENT (DC), ALTERNATING CURRENT (AC), RESISTANCE, AND TEMPERATURE

Thinking classically again, we visualize a flow of billions of electrons through the circuit from the negative terminal of the battery to the positive terminal. But the electrons have a negative charge. If negative charges flow in one direction, it is the same electrically as what would happen if positive charges were to flow in the other direction, and so we say that a positive *direct current (DC)* effectively flows from the positive terminal of the battery through the circuit to the negative terminal (under the electrical pressure, i.e., the electrical *potential* described in Chapter 11, provided chemically as voltage by the battery). When electrical pressure is made to drive the current first in one direction and then the other direction, as happens when we use the voltage provided at wall outlets from far distant generators (using transformers and transmission lines in between), we say that we have an *alternating current (AC)*.

Because electron flow through *normal metals* involves *collisions* with impurities and other defects in the metal (which unlike the ions of the metal are typically not of uniform size and charge, or not on a regular lattice), the impurities and defects are *jostled*, and they in turn jostle the ions of the metal and cause vibrations throughout the entire crystalline lattice. We call the collective effect of these collisions that impede the flow of electrons *resistance*.

The introduction of this jostling of the lattice is called *heating*; and the degree to which the atoms are jostled around in the general location of their lattice positions is a measure of their *temperature*. (We could of course cause or increase the jostling, i.e., *heat* the metal, by letting it come into contact with more rapidly moving, *hotter,* molecules in the air of an oven, or by touching the metal with other *hotter* materials whose atoms are already jostling with even higher energy [i.e., at a higher temperature]. When the atoms are jostled sufficiently that the crystal lattice breaks down and the atoms find themselves continuously bounced into new locations, we say that the solid *melts* and becomes a liquid.)

Although we think of copper wire as being a good conductor, the electrical resistance of copper and other normal metal wires is high enough so that for the miles and miles of wire in electrical generators, transmission lines, and transformers, a great deal of electrical energy is lost as heat. Even if all of the defects could be eliminated, heating would still occur in metals because the normal jostling of the atoms at ambient (typical outdoor or indoor) temperatures distorts the crystal lattice of atoms enough that it causes collisions that interfere with electron conduction, in a similar manner to the interference caused by defects. Cooling would reduce the jostling and reduce this loss, but the cost of refrigeration to cool away the remaining losses would be prohibitive.

So it would seem that we are stuck with resistance and its associated energy losses, even with the best of metals. But nature has supplied a way to avoid the jostling and this heating: the quantum phenomenon of superconductivity, soon to be described in Chapter 19.

Part Five

QUANTUM WONDERS IN MATERIALS AND DEVICES, LARGE AND SMALL

Chapter 18

NANOTECHNOLOGY AND INTRODUCTION TO PART FIVE

So far in this book we have been operating in the realm of science: physics toward understanding the atom and developing the roots of chemistry. Now we enter the realm of application and invention: *applied* physics, chemistry and materials science, engineering, and metallurgy.

In this Part Five, I mention or describe some of the many materials and devices devised with the help of our knowledge of quantum mechanics (or simply understood after invention through our knowledge of quantum mechanics). I mainly address the more physics-based inventions, but I also include in Chapter 22 examples of recent developments using new forms of carbon in composite materials. These follow from the discussion of carbon and its compounds in Chapter 15. We consider *nanotubes, carbon-fiber composites,* and one "dream" application that utilizes these materials for flight.

While materials have been developed and used in many ways since ancient times and much invention has occurred even without complete understanding, further invention has often been spurred by both modern classical concepts and quantum mechanics. Indeed, as noted in the preface to this book, fully one third of the US economy involves products based on quantum mechanics.[1] In the applications of modern chemistry and biology we have a host of plastics, polymers, coatings, paints, cosmetics, and medicines too numerous to mention. But the influence of quantum mechanics has been most directly felt in areas of the physics of materials and electronics, and it is in these more physically based areas that we concentrate here. As background to this discussion I described in Chapters 16 and 17 the makeup of solids and their electrical properties.

One fascinating *optical quantum device* with a multitude of applications, the laser, was presented in Chapter 4. We will discuss its application in fusion power generation in Chapter 20. This device and most of the rest of the modern wonders that we will consider are in some way electrical and involve the conduction of electricity through metals or the alloys of metals. In Chapter 19 we discuss *superconductivity*, a fantastic quantum mechanism for electrical conduction that has been used to achieve *something akin to perpetual motion* (without most of us knowing about it, in a modern device that we have used extensively).

Superconducting devices usually operate to either detect very small or weak magnetic fields or create very large and intense ones. One application under development that uses large-volume high magnetic fields is the *superconducting maglev train*, shown in Figure 19.1. (I describe more of this development in Chapter 19.) In Chapter 21, I discuss *magnetism, modern magnetic materials and their applications, and what is meant by a "high" magnetic field*, a field so strong that it can levitate a frog.

Most of us are familiar with modern inventions in *electronics*, and most of these inventions involve *semiconductors*. But few of us have the faintest clue as to how these electronic devices operate, or how one manages to place *billions of electronic circuit elements on a chip the size of a small postage stamp* (so that, for example, computers can store and rapidly process enormous amounts of information). In Chapter 23, I discuss how this is done.

In Chapter 24, we return to superconductivity to consider large-scale applications of superconductors; for example, in the search for the Higgs boson and in efforts underway to develop new sources of electric power and to improve the operational features and efficiencies of components for power generation and transmission on a scale to light even the largest of our cities. It is in this field of superconductors and superconducting devices that I have contributed over a span of some forty years as inventor, physicist, materials scientist, metallurgist, engineer, and project manager; and I describe here some of the inventions and developments that are part of my own experience.

Note, in many of the chapters that follow, we will often be examining what has broadly been referred to as "nanotechnology," innovations that work with entities on the scale of the nanometer (one billionth of a meter), perhaps ten to a hundred times the size of atoms. I mention just two recent innovations here that fit that category but do not conveniently fit into the subject matter of those chapters. These are innovations that affect the strength and weight of solids.

In her article in *MIT Technology Review* of March/April 2015, Katherine Bourzac describes the groundbreaking work of Julia Greer at California Institute of Technology. Among her other accomplishments, Greer has created:

> a ceramic that is one of the strongest and lightest substances ever made. It's also not brittle. In a video Greer made, a cube of the material shudders a bit as a lab apparatus presses down hard on it, then [the cube] collapses. When the pressure is removed, it rises back up "like a wounded soldier."[2]

Julie Shapiro, in *Time* magazine of November 2015, describes in a page labeled "Breakthrough: 'A Metal That's (Almost) Lighter Than Air'":

> The world's lightest metal contains hardly any metal at all—in fact, it is made of 99.99% air. Called a microlattice, the material is a three-dimensional grid of tiny tubes that is up to 100 times as light as Styrofoam. And it is poised to revolutionize the way we fly.[3]

I welcome you to the wonders of Part Five.

Chapter 19

SUPERCONDUCTORS I—DEFINITION, AND APPLICATIONS IN TRANSPORTATION, MEDICINE, AND COMPUTING

Superconductors have two special properties that allow them to do things that normal conductors cannot do. Both result from phenomena that can only be explained through quantum mechanics. One relates to a quantization of magnetic fields. The second is responsible for the name "superconductors," for these materials have the capability of carrying electrical current totally without any electrical resistance.

SUPERCONDUCTIVITY

When cooled to sufficiently low temperatures, some metals, alloys of metals, and compounds suddenly lose *all* trace of electrical resistance, each at its own specific low *critical temperature* (T_c). This phenomenon, called *superconductivity*, was discovered by the Dutch scientist Kamerling Onnes in 1911 in mercury cooled in liquid helium. Physicists John Bardeen, Leon Cooper, and Robert Schrieffer, in 1972, received the Nobel Prize in Physics for work done in 1957: "*For their jointly developed theory of superconductivity, usually called the BCS theory.*"[1] The theory has been of considerable help in guiding the search for superconductors with higher critical temperatures, though new high-temperature[2] superconductors (to be described) have shown a need for additional theory.

Below the critical temperature, each electron of plus ½ spin is found (in a complex way) to "*pair up*" with another electron of minus ½ spin to form a *quasiparticle* with a net spin of zero. Particles or quasiparticles with zero spin, unlike spin ½ particles, experience no exclusion. They tend to collectively occupy a single lowest-energy superconducting state. But the entire sea of electrons must be in this state at the same time. And any collision (even of one quasiparticle with an impurity that distorts the lattice) has to be of sufficient energy to break apart the entire collection of paired particles. (It's as if an army of policemen with locked arms was to march down a street. It would be a difficult thing to break through all of those locked arms simultaneously to disrupt the march.) And so the entire collection of electrons remains impervious to collisions with impurities—and impervious to the lattice oscillations that normally contribute to electrical resistance, so long, of course, as the superconductor is kept below its critical temperature and below a certain *critical current* level.[3] In this superconducting state there is no resistance, that is, no heat is generated, and only a relatively small amount of refrigeration is needed to remove any heat that leaks into the cold region from the outside world (through a Thermos-bottle-like type of thermal insulation).

MAGLEV—MAGNETICALLY LEVITATED TRAINS

One of the more exciting commercial developments of superconductivity is *magnetically levitated trains*. Trains using permanent magnets and more conventional electromagnet technology have also been under development, and some are already operational.

The Japanese have been developing the world's most advanced *superconducting maglev* train as a successor to their famous *Shinkansen* (bullet train). One of the maglev trains is shown in Figure 19.1 on its Yamanashi test track. Another of their trains (in 2003) set a world's record for high-speed trains at 361 miles per hour (581 km/h).

The "track" is not really a track in the conventional sense. The train has wheels for operation at slow speeds. But once the train gets moving, the strong constant magnetic fields of superconducting magnets under the cars of the train *induce* electrical currents in an electrically conducting *guideway*. These induced currents in turn produce magnetic fields that repel the magnets on the train, and at a sufficient speed this repulsion is strong enough to lift the train, *wheels-free,* above the floor of the guideway. Similar forces keep it centered *laterally within* the guideway, and a linear electric motor couples magnetically to the guideway to propel the train.

Fig. 19.1. Superconducting magnetically levitated (maglev) five-car train on the test "track" in Yamanashi, Japan, September 5, 2013. (Image from *Wikipedia* Creative Commons; file: SCMaglev Series L0.jpg; author: Saruno Hirobano. Licensed under CC BY-SA 3.0.)

Both the magnetic levitation and magnetic propulsion allow for a smoother, quieter ride, avoiding the sensitivity to conventional track misalignment experienced with more conventional trains. Alignment, and associated track maintenance and cost, would be a major issue were conventional trains to run at 300 miles per hour.

In 2011, the Japanese gave the go-ahead[4] to begin construction in 2014 on a line initially between Tokyo and Nagoya and to enter commercial maglev operation in 2025. Later extensions to Osaka are to be completed in 2045. The estimated travel time for the 272-mile (438-km) Tokyo to Osaka run is just 67 minutes, making it faster by 44 percent than the rapid runs of the present Japanese bullet trains.

MRI—MAGNETIC RESONANCE IMAGING

As Rosenblum and Kuttner nicely summarize, MRI (for which the Nobel Prize in Physics was awarded in 2009) "is made possible by the coming together of the quantum phenomena responsible for nuclear magnetic resonance (NMR), superconductivity, and the transistor."[5] NMR is involved because the quantum spin states of the protons in the nucleus of certain atoms in our bodies are sensitive to the conditions of surrounding tissue, and these are probed using resonant radio waves. The fundamentals of semiconductor operation, and the powerful computers that are involved in producing images from these measurements, are described in Chapters 17 and 23. And the role of superconductivity in providing the rock-solid high steady magnetic fields needed for MRI is described here.

The most successful present commercial use of superconductors is for the MRI scans used in medical diagnostic imaging.[6] Patients slide into large tubes surrounded by high-field solenoidal magnets within thermally insulating vacuum jackets. These magnets produce a highly uniform base field in the region of the body to be imaged. The magnets are wound from miles of niobium-titanium alloy superconductor wire, and they operate immersed in liquid helium at –452 degrees Fahrenheit (–269 degrees Celsius). At this temperature, air and all other substances are frozen solid.

(The Celsius scale is particularly convenient for the temperatures that we normally experience, with zero degrees set as the melting point of ice and 100 degrees set as the boiling point of water. When considering very low temperatures, however, a *Kelvin* scale is used. Each degree of temperature is the same size as for the Celsius scale, but the absolute zero of temperature, the lowest temperature theoretically possible, which occurs at –273 degrees Celsius, is defined as the beginning of the Kelvin scale, zero degrees Kelvin. So, one measures temperature upward in Kelvin degrees from absolute zero. The critical temperature of niobium-titanium is at about 9 Kelvin, and magnets made using this alloy are usually operated immersed in liquid helium, which boils at 4.2 Kelvin at standard atmospheric pressure. On the Kelvin scale, ice melts at 273 degrees, so you get the idea: 4.2 Kelvin is pretty cold, everything is frozen solid there except liquid helium!)

An MRI magnet is energized by applying a voltage that drives the increase in the current through its windings to create a uniform, high (~2 tesla) magnetic field[7] in the bore (patient region) within the vacuum-jacketed magnet. Then a heated piece of superconductor across the terminals of the magnet is allowed to cool, forming a superconducting *short*, and all external electrical connections are removed. Because the magnet and its short are entirely superconducting, with almost-perfect superconducting joints, there is essentially no resistance in the electrical circuit, and the current continues to circulate in a very close approximation to perpetual motion. The magnet thus maintains an essentially fixed precise magnetic field. (The field will actually drop very slightly, by less than a millionth of a percent per year. This is because of the small resistance in the joints at the short and between wire sections of the magnet which convert some of the energy stored in the magnetic field to heat.)

Except for this very small amount of heating produced by resistance in the joints between sections of superconductor and connections to the short, no heat is produced in the superconducting winding. But heat does try to leak through the vacuum jacket from the outside. Most of this heat is intercepted using a *cryocooler*, a small refrigerator that you may hear making a continuous clicking, thudding, or pulsed hissing sound in the background.[8] The small amount of heat that does get through the vacuum jacket and past the refrigeration slowly boils away the liquid helium (which, like water when it is boiled, holds temperature at its boiling point until it is entirely boiled away). In modern MRI systems, the refill period is once every three years.

OTHER MEDICAL, SCIENTIFIC, AND COMMERCIAL APPLICATIONS

To date, superconductors have found major commercial application only in *MRI* magnet systems, as just described. However, they have also been used to create highly attracting magnetic fields for the removal of (dark-colored) magnetic impurities in the purification of clays for *the manufacture of white fine china dishware*. Superconducting quantum interference device (SQUID) magnetometers[9] have been developed for the *ultra-*

precise measurement of magnetic fields toward both scientific and commercial applications. Though the sale of magnetometers is a relatively small business, the value of ores discovered using SQUID magnetometers is easily in the range of tens of billions of dollars.

Biomedical applications include neuroscience[10] and cancer detection.[11]

One recent development involving SQUIDs is worthy of special mention. *A safe ten-minute, noninvasive method, much more accurate than and superior to electrocardiograms, for the quick early detection of heart problems* has been developed by Cardiomag Imaging in Latham, New York. The technique uses an array of SQUIDs to measure the very weak magnetic fields produced by the small electrical currents that drive the muscles of the heart. This method has FDA clearance and other regulatory approvals, but not yet the necessary payment reimbursement codes from insurance companies. Cardiomag is arranging the necessary funding to obtain such reimbursement and launch into the large-scale manufacture and marketing of their imaging systems. Once the necessary financing is received for this purpose, the broadly based use of this innovative advancement is projected to save many hundreds of thousands of lives and avoid expenditures of tens of billions of dollars in annual healthcare costs.[12]

Superconducting electronics have aided our search of the cosmos, both in support of the Atacama Pathfinder Experiment and a the Atacama Large Millimeter Array radio telescope activities in the high Atacama desert of Chile.[13] Finally, I note that developments have been underway off and on since the 1960s to examine the possible uses of superconductors to minimize heating as a major obstacle in making *faster and/or more compact computers*. I link to one update in August 2015[14] and further note the December 2014 announcement (of the Intelligence Advanced Projects Activity) of a multiyear project to develop a superconducting computer.[15] The most recent and most exciting of these developments involves *quantum computing*, using Josephson junctions (the key superconducting element involved in SQUIDS) to produce quantum-linked storage bits called *qubits*. I report more on this development in Chapter 8, on quantum computing, and in its related Appendix C.

Chapter 20

FUSION FOR ELECTRICAL POWER, AND LASERS ALSO FOR DEFENSE

Throughout this book, we are concerned mainly with quantum mechanics as applied to the electrons surrounding the nucleus and the associated electronic "chemical" (rather than nuclear) properties of the elements. But in this chapter we discuss several large-scale fusion power approaches that involve the nucleus of the atom. Here we deal with quantum-based nuclear processes and quantum-based technologies that are employed to enable the fusion process. And, along with presenting the laser as a tool for fusion, I also mention several laser-based military applications.

Nuclear processes such as fission and fusion are understood through a quantum mechanics that considers what happens inside of the nuclei of atoms. In these processes we deal with *isotopes*. Atoms having the same atomic number (that is, the same number of protons in their nuclei) but having different numbers of neutrons in their nuclei are different isotopes. Most elements can be found in several different isotopic forms.

FISSION AND FISSION REACTORS

It is the fission (splitting) of some isotopes of the heavier elements (such as the isotope of uranium, U235, Z = 92 with the atomic weight of 235 protons and neutrons) into atoms of lighter elements that produces what has generally been called atomic *energy*.

Isotopes that undergo a spontaneous fission are called "radioactive." It is the carefully controlled stimulation of this radioactive release of energy that is at work in our present-day nuclear power plants.

(The uncontrolled "chain-reaction" of the fission of one isotope to release neutrons that trigger the fission of more isotopes produces the enormous energy of the "atomic bomb.") Exaggerated fear of such a runaway situation in fission reactors [such as occurred at Chernobyl and Fukushima] has unfortunately curtailed the use of this present lowest-cost, carbon-emissions-free, present mode [see * below] for generating electrical power.

Those were very unfortunate accidents, but modern reactors today may be considered to be safe. And work is underway to develop inherently even safer and smaller and less-expensive fission reactors. For example, Terrestrial Energy in Mississauga, Ontario, Canada, is designing a reactor that uses molten salt, rather than water, as a coolant, essentially making the reactor meltdown-proof.[1])

*In a paper to be quoted later in this chapter,[2] Lev Grossman shows the results of a study giving the energy produced per dollar to build and maintain the power plant: fusion (projected for the future) at 27 wins over (present-day) gas at 5, coal at 11, and fission at 16.

FUSION TO LIGHT CITIES, THE "HOLY GRAIL" OF ENERGY SOURCES

As noted above, fusion is the same process that releases the energy that heats our sun and other stars. Because its fuel is or is derived from isotopes of hydrogen that are found in vast quantities (compared to what will be needed) in either freshwater or seawater, there is an essentially limitless supply of fuel. Enormous as the energy may be in an atomic bomb, it is small compared to the fusion energy that is released in a hydrogen bomb. But since the fusion reaction tends to stop in a fusion reactor if there are any problems, and there are no radioactive "spent" nuclear fuels to store or dispose of, nuclear fusion is an inherently safe, carbon-emissions-free, source of energy for our planet. These factors, plus its ratio of power delivered to relative cost (as noted above), make nuclear fusion the "holy grail" of energy sources.

In fusion, the nuclei of atoms of lighter elements fuse together to make heavier elements, and they give off a tremendous amount of energy in the process. Tritium is an isotope of hydrogen, $Z = 1$, having the same single proton and single electron as hydrogen, that is, behaving "chemically" the same and occasioning no new place in the periodic table, but having two neutrons in its nucleus in addition to the one proton. And so, since a neutron is about as heavy as a proton, the tritium isotope is three times as heavy as hydrogen. Deuterium, known as "heavy hydrogen," has one neutron in addition to the proton in its nucleus, and so it is twice as heavy as hydrogen.

A very large amount of energy is released as each tritium nucleus combines by fusion with a deuterium nucleus to form a nucleus of helium, giving off a highly energetic extra neutron in the process. Tritium can be produced in sufficient quantity for a fuel as these neutrons combine with the nuclei of lithium in a surrounding "blanket" that captures the energy and prevents the neutrons from irradiating and destroying its surroundings. So the reaction of the neutrons with the blanket generates the reactor's own tritium fuel. As noted above, deuterium is present in essentially limitless quantity in seawater, so getting enough deuterium is not a problem.

And the process is safe. In the tokomak, a loss in the magnetic confinement of the fusion plasma abruptly shuts off the fusion process, which needs the confinement to be sustained. And because for laser fusion tritium and deuterium fuel is provided in tiny pellet-sized doses, only a limited amount of fuel is available for each laser pulse. If anything goes wrong, the pellets aren't supplied or the energy delivered by the lasers is insufficient to ignite the reaction.

Fusion in a Gigantic Superconducting Torus, the Tokomak

The leading approach for fusion power, the tokomak, relies on the quantum phenomenon of superconductivity for effective operation. A *plasma* of hydrogen isotopes is heated and contained in the magnetic field of a "tokomak" (so named by the Russians who first explored this toroidal [doughnut-shaped] field arrangement for magnetic confinement fusion). The magnetic field, which holds the hot plasma as if it were in a magnetic bottle, is provided by as few as six D-shaped *superconducting* coils that are stacked together with the straight parts of the D back to back so that the toroidal (doughnut-shaped) magnetic field rings the entire stack through the open center of the Ds. (Superconductors are described in Chapters 19, and 24.) Over sixteen megawatts of fusion power has already been generated using this arrangement, close to the power required to ignite the plasma. *Ignition* occurs when more energy is produced by the fusion reaction than is put into it to get the fusion process started. Were the windings of the tokomak not superconducting, economically producing the toroidal magnetic field of the tokomak would be a much harder and possibly impossible project. Take a look at the enormous power consumption and vast flow of cooling water already required for the very much smaller copper alloy Bitter magnet described in Chapter 21!)

The exploration of this more favored of the different types of fusion approaches is of such great scale and cost that it is being pursued in a joint effort by the more developed nations of the world. Construction of the next large experimental device, called ITER (International Thermonuclear Experimental Reactor), is already well underway in the south of France and is scheduled to begin operation in 2020 at an expected total cost of $20 billion. The construction of a demonstration reactor has been proposed to begin in 2024. The full-scale commercial tokomak reactor might stand fifty feet high, with an overall torus diameter of fifty feet. An update on the ITER project has been provided in the American Physical Society magazine, *Physics Today*.[3]

Various projects have been underway worldwide to explore alternative magnetic confinement fusion approaches. For example, it was announced in January 2016 that an alternative, stellarator, geometry of magnetic-confinement system, Wendelstein 7-X, has just been started up after eighteen years of construction. It has completed the first step, producing a plasma.[4] (The "Stellarator," a name coined to reflect the intent to create the conditions within a star, is like a torus, but the field windings are of such geometry as to cause the plasma to twist like a single braid in a multistrand hoop.)

Recent Designs for Nuclear Fusion at a Smaller Scale

Some other approaches to fusion that are being explored are described in an article in *Time* magazine of November 2015.[5] Lev Grossman reports on an innovative fusion development in a $500 million startup company called Tri Alpha Energy in Orange County, California. Grossman sets the context by explaining that this is one of a number of small commercial operations tackling the fusion problem, including efforts that he outlines as being pursued at General Fusion outside of Vancouver and Helion Energy in Redmond, Washington.

The challenge for fusion is getting the plasma hot enough and keeping it hot long enough. The Tri Alpha Energy reactor approach is described as having two canons "firing smoke rings, except that the smoke rings are hot plasma rings."[6] They are shot at each other at "just under a million kilometers per hour." The "violence of the collision of the two rings heats the combined plasma to ten million degrees Celsius and combines the rings into a plasma 80 centimeters across, shaped more or less like a football with a hole through it the long way, quietly spinning in place. . . . The plasma itself generates the field that confines it." In June the reactor proved able to hold its plasma for (a long) 5 milliseconds, indicating that a stable plasma had been created.

Gigantic Laser-Produced Nuclear Fusion

The laser is a quantum device, but that doesn't mean that it must be small. In a $3 billion experiment toward one *long-shot* application, 192 huge lasers were assembled to converge a pulse of simultaneous laser beams from many directions onto a pellet about the size of a peppercorn.[7] The objective was to use the power and coherence of lasers (as described in Chapter 4) to vaporize the pellet so rapidly that the expanding vapors compress and heat the pellet's tritium and deuterium contents to the enormous pressures and temperatures required to "*ignite*" controlled nuclear *fusion*.

The laser-fusion concept was first examined in the 1970s at KMS Industries in Ann Arbor, Michigan, and this examination was continued, starting in the 1980s, at the US government's Lawrence Livermore Laboratory (about an hour east of San Francisco). Construction of the laser-fusion National Ignition Facility was begun at LLL in 1997. The device occupies an area as large as three football fields.

In July 2012, this laser system delivered a whopping peak power of 500 trillion watts.[8] In September 2013, a laser shot produced fusion within the pellet that delivered more power than was driven into it, but this fusion energy was well short of the energy put into the entire system, as would be required to meet the goal of ignition. For the present time, at least, the machine is being used for other purposes.[9]

PROBABILITIES AND TIME FRAMES FOR FUSION POWER

Even if ignition is achieved, major engineering problems still need to be solved, and the construction and successful operation of a commercial fusion reactor by any approach is viewed as a *long shot* and probably would not take place before 2050 (though the aim of startup companies with their smaller-scale approaches is to do it much sooner). In any case, many argue that the prospect of safe, unlimited energy supply and the related prospect of a safer world (with less fighting over limited energy resources) makes it worth the development efforts that are underway.

LASERS IN DEFENSE APPLICATIONS

Not only does the generation and coherence of the laser beam as described in Chapter 4 allow it to have great power, it also allows the intensity of the beam to stay intact over long distances, rather than becoming spread out and diffuse, as occurs with ordinary light. This property has made the laser attractive for laser-guided weaponry, and as a potential antimissile weapon, since intensive bursts of laser power could potentially be beamed out at the speed of light over long distances to destroy incoming missiles. The US Department of Defense also uses the laser to examine the processes that would take place during the explosion of nuclear bombs or the nuclear warheads on missiles. This is a parallel use of the huge laser system at LLL that, as described above, was also targeted at developing unlimited fusion power toward a more peaceful world.

Chapter 21

MAGNETISM, MAGNETS, MAGNETIC MATERIALS, AND THEIR APPLICATIONS

Depending on how their electrons fill the available energy levels, and the degree to which they fill subshells, atoms or molecules of atoms may have net magnetic moments. (Remember from Chapter 12 that electrons have magnetic moments that can be thought of as resembling tiny bar magnets, and the moments result from either (or both) the electron's intrinsic spin or the magnetic component of their spatial-state angular momentum.) These moments tend to be oriented toward or opposing a magnetic field, depending on the plus or minus spin and/or angular-momentum state that the electron occupies.

DIAMAGNETISM

Those substances whose atoms have no net magnetic moment are called *diamagnetic substances*. (Most people would call these substances nonmagnetic). In fact, all substances will have a small diamagnetic component to their magnetism. It arises from the induced response of the electrons in the quantum spatial states of their atoms as any magnetic field change is applied to them. Classically, one can think of small electrical currents induced within the atom that produce small magnetic fields that oppose the field change. Because these small induced atomic magnetic fields oppose the field that is applied, the atoms are repelled by field change, and so diamagnetic substances are very weakly repelled if a magnetic field is changed on them.[1]

Most substances are diamagnetic. Living things tend to be diamagnetic. And so it is that scientists have even been able to repel and levitate a frog and other small biological items using the magnetic field produced in one of the most powerful magnets in the world.

MAGNETS AND MAGNETIC FIELDS

In the paragraphs below, I start with a description of "frog levitation" as a mechanism for providing a sense of the magnitudes of magnetic fields. I also mention various devices that produce or use magnetic fields, and I describe materials that produce or duct magnetic fields around.

The source of the levitation field for the frog mentioned above is a 1.25-inch bore 16-tesla Bitter magnet located at the Nijmegen High Field Magnet Laboratory in the Netherlands. (*Bore* is the available cylindrical open center space in which a patient or specimen may be placed.) Sixteen tesla[2] is an extremely high magnetic field. It is approximately 320,000 times the earth's magnetic field; approximately 100 times the magnetic field of household permanent magnets such as those that you stick on your refrigerator doors; approximately 10 times the field of the strongest of permanent magnets; approximately 10 times the field conducted through ferromagnetic materials in the iron magnetic circuits of large transformers, motors, and generators; and approximately 10 times the field inside of the large bore tube of the superconducting magnet that you slide inside of for MRI medical diagnostic imaging (as discussed in Chapter 19).

In the operation of a Bitter magnet, enormous electrical currents are forced through a flat, mechanically strong yet still good-electrically-conducting alloy (that is wound on edge like a Slinky in a tight spiral around the bore) to create the magnetic field. There is some resistance to current flow in a copper or alloy winding, and this resistance produces heat, so coolant is also forced through the winding to remove the heat. This type of magnet is named after Francis Bitter, who developed it to allow a cooling of electromagnets. In 1938, Bitter established a (later called "national") magnet laboratory at MIT and achieved a record steady magnetic field of 10 tesla. Prior to the Bitter magnet, there was no good way to cool electromagnets, and the magnetic fields that they could produce were limited. (During World War II, Bitter worked on finding ways to demagnetize British ships to protect them against German field-sensing mines. This work Bitter called "Degaussing the Fleet."[3])

The world's highest sustained (rather than pulsed) magnetic field is available for experimental work at the (present) US National High Magnetic Field Laboratory in Tallahassee, Florida. This field is produced by a *hybrid* magnet system consisting of an outer superconducting solenoid[4] that provides an 11.5-tesla field boost for an inner 33.5-tesla Bitter magnet, to produce a total of 45 tesla in a 1.25-inch bore. Huge currents through the resistive Bitter magnet continuously dissipate 33 million watts (megawatts) of power as heat. (This power would light 33,000 average homes!) A water flow of 4,000 gallons per minute is required to cool the magnet. By contrast, the superconducting solenoid requires only a few hundred watts from standard outlets: (1) to run the refrigerators that help to keep the superconductor cold (in a manner similar to that described for MRI magnet systems in Chapter 19) and (2) to supply the power needed (after conversion from AC to DC) to drive sufficient current through the solenoid to supply the energy that is stored as the magnetic field.

PARAMAGNETISM

In many instances, diamagnetism is overcome by the direct pull of a magnetic field on the spin and spatial-state magnetic moments of the atoms. Atoms or molecules may have net magnetic moments that reluctantly align with and are attracted into an applied magnetic field (reluctantly because the magnetic moments are thermally jostled at normal room temperatures toward random orientation). These substances are said to be *paramagnetic substances*. (They are only weakly magnetic, and most people would also call these substances nonmagnetic). For example, oxygen is paramagnetic. As a demonstration of this, liquid oxygen can be held in suspension between the jaws of a strong horseshoe magnet. (Note that it is the absorption of particular wavelengths of light in paramagnetic oxygen trapped in glacial ice that gives the ice its blue color.)

FERROMAGNETISM

In a few materials (iron, cobalt, and nickel, for example), the magnetic moments of large numbers of atoms or molecules will spontaneously align to create very strong collective magnetic moments within regions of the material called *domains*. In the absence of a magnetic field, these domains arrange themselves so as to duct the magnetic field of these moments around in closed magnetic circuits. However, even in a weak applied magnetic field in what are called "soft" magnetic substances, these magnetic moments can become fully aligned to create a single domain and a large associated magnetic moment that is strongly attracted into the magnetic field. These *ferromagnetic substances* are what people normally think of as "*magnetic*," and we say that these materials are *magnetized* when the single domain is created.

Ferromagnetism is a special situation that depends not only on the magnetic properties of the atoms in a solid substance but also on the spacing and magnetic connection between the atoms. The details are explained as follows.

> Theoretically, for ferromagnetism to occur, the spatial states of the outermost electrons of adjacent atoms in a solid substance must have a substantial overlap, while at the same time these states must show a limited probability for the electron to reside near the nucleus. Their nuclei must also be spaced appropriately apart. The d and f states are apparently particularly suitable to these requirements. (The probability clouds for these states may somewhat resemble those shown for the d and f states of hydrogen in Figure 3.8.) And so we can understand that iron, cobalt, and nickel in the fourth row of Table IV in Chapter 13 or Table B.2 in Appendix B (with six, seven, and eight outer 3d electrons, respectively) and gadolinium and dysprosium in the sixth row (each with outer 4f electrons) all are ferromagnetic. Interestingly, while manganese (with five outer 3d electrons) by itself is paramagnetic, not ferromagnetic, a couple of compounds of manganese that force the manganese atoms to be slightly farther apart than they are in the pure metal are ferromagnetic, apparently reaching the right separation for the manganese d states.

APPLICATIONS FOR FERROMAGNETIC MATERIALS

Just as copper or superconducting wires are used to duct electrical currents around electrical circuits, ferromagnetic materials are used to duct magnetic fields around magnetic circuits. Magnetic circuits are essential major components of electric motors, generators, and transformers. Since the transformer is an AC device that requires that the magnetic field alternate direction sixty times per second (in the United States and Canada, for example), its magnetic circuit must be made of the "soft" magnetic materials in which the domains are easily and quickly oriented to the continually changing field. Alloys of pure iron with small percentages of silicon have proven to be particularly good for this application.

Magnetic data storage uses magnetizable material to form a non-volatile memory. Data is "written" into the material (which is usually a surface coating) via heads that magnetize the material. The same heads may be capable of "reading" what is written by sensing the pattern of magnetic fields that are written there. This form of magnetic storage is in the "hard disks," in hard drives of computers, and in the stripes on our credit cards.

PERMANENT MAGNETS AND THEIR APPLICATIONS

Permanent magnets are made from ferromagnetic materials that tend to retain a single domain (or set of parallel domains) once their domain orientation is established. One primary way of accomplishing this is to create particles so small that they can contain only one domain, and then have these particles oriented in the presence of a magnetic field as the particles are pressed together to make a solid. Another way is to cause phase changes in solids that precipitate out tiny single-domain regions of one material composition (all oriented parallel to an applied magnetic field) within an overall surrounding matrix material of another composition. Permanent magnets are used, for example, in small motors, in devices for magnetically separating materials, in quadrupole magnets for scientific instrumentation, and for refrigerator magnets now popular for advertising or simply mounting notes or business cards. Rare-earth magnets are particularly important for high-field, low-weight applications of these materials. In this regard, neodymium iron boride is still the champ.

Chapter 22

GRAPHENE, NANOTUBES, AND ONE "DREAM" APPLICATION

I report here first on new forms of carbon with amazing properties that promise in time a host of inventions. While the carbon-to-carbon bonds can be understood from the quantum nature of the atom, these materials have additional properties predicted from the quantum mechanics of one- and two-dimensional materials. (I'll describe the predictions without getting into the theory.) I conclude with a success story of one application that extensively uses an advanced carbon-fiber composite technology that may someday incorporate some of these materials. It also uses a package of controls and communications equipment that relies heavily on the quantum devices that are described in the next chapter.

GRAPHENE AND RELATED FORMS (ALLOTROPES) OF CARBON

Graphene

Graphene is the basic structural configuration for many forms of carbon: some fullerenes, including nanotubes, and the more familiar graphite, charcoal, and soot. In graphene, carbon atoms are at the vertices of a honeycomb-like arrangement with chemical bonds between them. The bonding carbon atoms form a sheet of connected hexagons all in a single plane, as shown in the scanning probe microscope image of Figure 22.1(a). (Note that boron nitride and other compounds can also be formed in monolayer sheets.)

The effective thickness of the graphene sheet is 3.4×10^{-10} meters, that is, 0.34 nanometers, where a nanometer is one billionth of a meter (about one millionth the thickness of a dime. Graphene is one hundred times stronger than the comparable thickness of a (hypothetical) sheet of the strongest high-carbon steel. The spacing between carbon atoms in the hexagons of the sheet is 0.14 nanometers, about the same as the spacing between the densely packed carbon atoms in diamond (a measure that is used for the effective diameter of the carbon atom, as shown in Table D.1.) So graphene is very compact in the plane of the sheet. Note that its thickness (given above) is about 2.5 times the carbon-to-carbon spacing. (The stability and strength of graphene, and that of the carbon nanotubes to be described below, are due to carbon's sp^2 hybrid of s and p spatial states [bonding as described in Chapter 15].)

Fig. 22.1. (a) Graphene magnified approximately twenty million times in this surface probe image. (Image from *Wikipedia* Creative Commons; file:Graphene SPM.jpg; author: US Army Materiel Command. Licensed under CC BY 2.0.) (b) Model of buckminsterfullerene at a similar scale. (Image from *Wikipedia* Creative Commons; file:C60 Buckyball.gif; author: Saumitra R. Mehrotra and Gerhard Klimeck. Licensed under CC BY-SA 3.0.) (c) A carbon nanotube with a nanobud at similar scale. (Image from *Wikipedia* Creative Commons; File:Nanobud.jpg; author: Arkady Krasheninnikov. Licensed under CC BY-SA 3.0.)

Graphene had been observed under electron microscopes in 1962, but it was not studied further.[1] However, as a result of its rediscovery in 2004, Andre Geim and Konstantin Novoselov in 2010 received the Nobel Prize in Physics *"for groundbreaking experiments regarding the two-dimensional material graphene."*

Graphene, per pound, is the strongest material ever tested. As mentioned in Geim's Nobel announcement, a graphene hammock one meter square that would support a 4 kilogram (8.8 lb.) cat would weigh less than one of the cat's whiskers.[2]

Wikipedia notes that by the end of 2014, the global market for graphene had reportedly reached just $9 million, mainly to the battery, energy, semiconductor, and electronics industries.[3] This is small compared to the billions that are apparently being spent to search out new applications for this substance (see below). To get a sense of the effort, note that a recent publication for graduate students and professionals entering the field uses some 450 pages just to catalogue and briefly describe work as of 2013 relating to this substance and its potential applications.[4]

In a 2014 article in *Scientific American*, Geim describes graphene as one component in the fabrication of stacks or sandwiches with other monolayer materials, including boron nitride and molybdenum or tungsten disulfides.[5] These composites can have special properties. He suggested possibly even ambient temperature superconductivity, but he remarked that as yet no "killer" applications have emerged (despite that graphene can be manufactured in sheets of hundreds of square meters). But perhaps that is about to change.

In a late 2015 article titled "Introducing the Micro-Super-Capacitor: Laser Etched Graphene Brings Moore's Law to Energy Storage," Maher El-Kady and Richard Kaner describe the work of a group at UCLA toward what seems a possible breakthrough technology.[6] Capacitors, the charge-storing components of electronic circuits, are the only circuit elements (along with batteries) that have not kept pace with the ongoing miniaturization that has allowed the development of small modern user-friendly electronic devices. The authors describe a graphene-based two-dimensional approach that would allow the integration of miniature, flexible, high-energy capacitors into modern solid-state electronic devices. The group is also developing battery/capacitor hybrids, with the aim of reducing the size of the battery as well. Commercial applications are now being explored by Nanotech Energy, a Los Angeles–based startup.

Katherine Bourzac[7] describes the use of graphene as a substrate layer upon which other electronic materials can be thinly deposited to make flexible devices, including liquid crystal displays (LCDs) that have flexed a thousand times without degradation. (The displays on our cell phones are LCDs.)

In its pure form, graphene has an exceptionally high electrical conductivity. While this is very good for some applications, it can be a problem for others. John Pavlus describes an effort at IBM ($3 billion in 2014) to explore the use of new materials, primarily graphene, to go beyond our present silicon-based semiconductor technology. "Graphene transistors have been built that operate 100s to 1,000s of times faster than top silicon devices, at reasonable power density, and below 5nm where silicon goes quantum."[8] ("Goes quantum" refers to being on such a small scale that the wavefunction of electrons in neighboring parts of the circuit begin to significantly overlap in a way that effectively shorts across circuit elements.) But graphene lacks a band gap (see Chapter 17) and so behaves as a metal rather than as a semiconductor. It can't turn off a current as a transistor does, and so it cannot encode digital logic. (A transistor current is turned on and off to physically store a bit of information, a 1 or a 0, as described in Chapter 8.) Pavlus notes that carbon nanotubes (to be described below) *can* have a small band gap and be semiconducting. Individual tubes show a fivefold improvement over silicon. But they are fragile and disturbances can remove the band gap.

Note that graphene is the only solid form of carbon in which every atom is available for reaction from two sides. It has a very high opacity (an ability to absorb radiation). And, as a monolayer, its properties are extremely sensitive to its surroundings. Performance can be degraded by impurities, but this opens possibilities for its use in sensors.

Graphite has been found in nature and used for various purposes as early as 4000 BCE[9] It is composed of graphene sheets stacked one atop the other, but slightly shifted so that the carbon atoms in the second sheet lie above the middle of the hexagons in the first sheet, the carbon atoms of the third sheet line up over the carbon atoms in the first sheet, and so on. Because the bonds between the sheets are relatively weak, the sheets find it easy to slide over one another. That is why graphite is "slippery" and sheets or groups of sheets can rub off, a property that has been conveniently exploited in the manufacture of pencils.

Fullerene

Fullerenes are a family of geometric structures having a monolayer surface of atoms (or molecules) somewhat like that shown for graphene. One of these structures is the *"buckyball,"* which has the type of configuration shown in the lines or seams of a soccer ball or the support structure of the large spherical geodesic dome at the entrance to Disney World's EPCOT Center. Another geometric form of the fullerene is the *nanotube*, to be described just ahead.

Figure 22.1(b) shows a model of the buckminsterfullerene C_{60}, prepared first in the laboratory and named "in homage of" Buckminster Fuller, who first started constructing the geodesic domes that this structure resembles. In $C_{60,}$ carbon atoms are at the vertices of "rings" of hexagons surrounding rings of pentagons (rather than having only the rings of hexagons, as in graphene). There are sixty carbon atoms in the structure. The spacing between the center of one ring and the center of the next is, on average, 0.14 nanometers, about the carbon-to-carbon length described for graphene.

The buckyball type of fullerene has also been produced with complexes of carbon and other atoms at the vertices. Various types of buckyballs have found application in the medical field for gene and drug delivery and in contrast agents for x-Ray and MRI medical diagnostic imaging.

Although first prepared in the laboratory in 1985, fullerenes have since been detected in nature and even in outer space. According to astronomer Letizia Stanghellini, "It's possible that buckyballs from outer space provided seeds for life on earth."[10]

Carbon Nanotube

The *carbon nanotube* (CNT) can be grown as a tube by various means in the laboratory, but it is also found in less-regular forms in ordinary flames produced by burning ethylene, benzene, and methane and in soot in indoor and outdoor air. The CNT can be thought of as a very long strip of graphene that has rolled itself about its long axis to form a tubular, seamless monolayer. The rows of hexagonal rings can just circle the tube's long axis or spiral around it. There are many variations beyond these structures, including, for example, a nanotube within a nanotube within a nanotube. Depending on the structures and variations, different electrical, thermal, optical, tensile, and compressive properties are obtained.

Figure 22.1(c) shows a model of a short section of a single carbon nanotube of the "ring circling" type with a fullerene "bud" attached (via covalent bonds). These buds may prevent slippage between nanotubes to improve the mechanical properties of bundles of nanotubes in the high-strength composite structures to be discussed just ahead.

Publication of a "discovery" of nanotubes by Sumio Iijima of NEC (Previously Nippon Electric Company) in 1991 produced a "flurry of excitement" about this new material, though a string of observations by investigators in various countries had been reported earlier, starting with work in the Soviet Union by Radushkevich and Lukyanovich published in the *Soviet Journal of Physical Chemistry* (in Russian) in 1952. Nanotubes have also been found in nature, in charcoal, and in soot.

Carbon nanotubes have been made as narrow as one nanometer in diameter ("nanotube" is fitting), the span of about seven carbon atoms. Like the buckyballs, they aren't visible without the aid of an electron microscope. But these tubes can be made millions of times longer than their diameters, so we have tubes perhaps ten atom widths in diameter extending to lengths on the order of the thickness of a dime. Some nanotubes have been constructed with length-to-diameter ratios of 132,000,000-to-1, a larger ratio then for any other material.[11]

Though half as strong as graphene, individual carbon nanotubes are still three hundred times as strong as high-carbon steels, and they will stretch by up to 5 percent rather than break under excessive strain (unlike graphene, which is a bit brittle). Most synthesized CNT arrays are hydrophobic (they repel water), but with the application of a low voltage they can become hydrophilic (attract water). Carbon nanotubes are frequently referred to as one-dimensional electrical conductors. Depending on how the rows of carbon rings circle or spiral around the long axis of the tube, the nanotube can be either electrically conducting or electrically semiconducting. The theoretical maximum conductance of a single-walled tube is described as resulting from the tube acting as "a ballistic quantum channel."[12] (A theory-based claim of "intrinsic superconductivity" is in dispute.) In theory, the conducting nanotube can carry a current density one thousand times that of metals such as copper. And at normal room temperatures, the CNT is expected to conduct heat along its long axis almost ten times better than copper, but transverse to that long axis the CNT is a good thermal insulator.

These properties have suggested a host of potential applications. I mention a few here. Note that a high current capacity is desirable for many applications, and a CNT-copper composite has been shown to carry one hundred times the current of pure copper or gold. Coating military aircraft with radar-absorbing CNTs may enhance their stealth capabilities. Toward computing, a nanotube-integrated memory circuit was made as early as 2004, but regulating the conductivity of the nanotube proved to be difficult. (From my experience, it would seem that the exceptional heat conduction of the CNT could help to solve one of the biggest problems limiting the compactness and capacity of computers: removing the heat produced in the switching of the billions of circuit elements in the processor and memory components of the computer.) CNTs are being looked at for improved electrodes in lithium batteries. Solar cells are being developed using a combination of buckyballs and CNTs, the former to trap electrons and the latter to conduct them away to deliver electrical power. CNTs have also been used as tips for scientific-force microscope probes and in medicine to provide a scaffolding for bone growth. And, in what might be an inadvertent and outmoded use, we note that CNTs have been found in Damascus steel from the seventeenth century, perhaps accounting for the legendary strength of the swords that were made from that material.[13]

Nanotubes are just one of a number of structures involved in the development of devices on a nanoscale. The inherent strength of CNTs makes them particularly attractive for this type of application. One experimental use for the nanotubes is for nanoscale bearings, where one nanotube, repelled to the center inside of a larger-diameter nanotube, can rotate essentially without friction. This property has been used to create the world's smallest rotational motor.[14]

Finally, CNTs may be useful in helping to solve the biggest problem related to solar and wind power: that these sources generate power intermittently and at times when it may not be needed. Ways are sought to store the energy produced during the "down" periods, so that it can be used when and where it is needed. One method is to use the electrical power produced from these sources to electrolyze water into hydrogen and oxygen, so that the hydrogen can be stored and transported. For automobiles in particular, safe storage at ambient temperatures is desired. One way to achieve this is to have the hydrogen molecules or atoms attach themselves to the surface of solid materials in a way that lets them detach for direct use in combustion engines or for conversion back, via fuel cells, into electrical energy to power electric cars. CNTs would offer an enormous surface area (per pound of material) for this kind of attachment. For this and many of the applications that I have mentioned, the challenge may be to produce CNTs of sufficiently high purity at sufficiently low cost.

Current applications mainly use bulk carbon nanotubes (masses of unorganized fragments of nanotubes) added into carbon-fiber composites for improvements in mechanical, electrical, and thermal properties of the bulk product. Composites containing CNTs have been used, for example, for added lightweight strength in bike components. (Note that carbon-fiber composites, even without CNTs, already have mechanical properties superior to the best of steels.)

LARGER-SCALE APPLICATIONS

The Space Elevator

Before leaving CNTs, I mention what I call a "way out" (pun and two meanings intended) potential application. (I don't expect you to take this too seriously.)

> Carbon nanotubes may be the only material strong and lightweight enough to enable the construction of a "space elevator." This device would allow payloads to be lifted up to an altitude of 22,000 miles (the altitude of stable Earth orbit) or beyond, there to be released for whatever missions may be involved. (This would save a lot of rocket fuel, and maybe rockets as well.)
>
> The elevator was first suggested by Konstantin Tsiolkovsky, one of the founding fathers of rocketry and astronautics,[15] in 1895 (in his "Speculations about Earth and Sky and on Vesta").[16] He suggested using a tower. However, since 1959, most ideas have involved tensile structures. And that is where the CNT comes in: the application depends on a lightweight and very strong cable, and the only possibly viable prospect to date is one constructed of carbon nanotubes.
>
> The elevator car would be hooked onto the middle of a cable between an orbiting "counterweight" in the sky (the "sky hook"?) and an anchor station somewhere on the equator. The counterweight would ride substantially above the 22,000-mile altitude, being thrown outward by the centrifugal force of the earth's rotation and kept from flying away into space by the equal inward centripetal downward pull of the cable, plus gravity. (The net of the downward gravitational force minus the upward centrifugal force is referred to as the force of the "apparent gravitational field.")[17] As the cable is paid out from the anchor station, the anchor (and the elevator hooked on far below it) would rise into space. To bring the elevator back down for a reloading, the anchor station would reel it in. The anchor, lower now but still well above the stable Earth orbit and pulling strongly upward, would be ready to lift the elevator again with its new payload.

It's time to come down to Earth. Well, almost.

Carbon-Fiber-Reinforced Polymer Composites[18]

Carbon fibers are about five thousand times the diameter of nanotubes, but still pretty small, about one-tenth the thickness of a human hair. They may be produced, for example, using yarns wound from rayon, from which all constituents except the carbon are driven off. In one approach for making the composite, carbon-fiber yarns can then be woven into fabrics, which are cut to size, layered in an appropriate mold, and impregnated with an epoxy resin. These composites have already been used in a host of applications, including sporting goods and automobiles. But at the turn of the century, another application was on the drawing boards at Boeing.

The Dreamliner[19]

Boeing Company may have staked its future on carbon-fiber-reinforced polymer (CFRP) technology in developing its model 787 Dreamliner, the most advanced new generation in commercial aircraft, as it entered commercial service in 2011.[20] Following tests performed earlier in relation to military aircraft design, wings and fuselage are no longer made from sheets of aluminum but instead use CFRPs to form a super-strong lightweight component. By making this plane lighter and stronger and by using advances in control systems, Boeing has been able to improve range and fuel efficiency (by 20 percent) without increasing the size of the aircraft, bucking the historical trend. The longest-range variant of the 787 can fly over nine thousand miles (New York to Hong Kong) without refueling. As of March 2016, Boeing had orders for 1,139 aircraft from 62 customers. (The development and flight testing of the 787 has been beautifully presented in an IMAX 3-D movie called *Legends of Flight*.[21])

But this is a competitive world. Responding to the threat that the Dreamliner posed to its business, Airbus in January 2015 introduced its A350, an aircraft of comparable capacity that is made 53 percent of composite structures, as opposed to Boeing's 50 percent. Variants of the two competing aircraft seat 225 to 350 passengers at a cost per aircraft ranging from $225 million to $356 million.[22]

Chapter 23

SEMICONDUCTORS AND ELECTRONIC APPLICATIONS

Those elements having properties in between those of metals and nonmetals are lightly shaded in the identical periodic Tables IV (in Chapter 13) and B.2 (in Appendix B). Among these semiconductors, also called semimetals, are silicon and germanium. The atoms of these solid elements, like those in most solids, tend to stack themselves over distances of millions of atoms (or more) in neat, even, three-dimensional crystalline arrays.

Quantum mechanics was utilized in the late 1940s and early 1950s (by Nobel Prize–winning physicists William Shockley and John Bardeen) to provide an understanding of semiconductors and the operation of semiconductor devices like the *diode* and the *transistor*. These basic devices are now present by the billions in the integrated circuits of *chips* that are at the operational core of most electronic devices. I describe next just how semiconductors work.

ELECTRICAL CONDUCTION IN UNDOPED SEMICONDUCTORS

Semiconductors were defined for the quantum view in Chapter 17 as solids in which the electrons either: (a) nearly fill the valence band below a band gap, (b) just fill a valence band below a relatively small band gap, or (c) just start to fill the states of the conduction band above the band gap. For each of these situations, electrical conduction can occur but is limited.

For semiconductors of type (a), the proximity of the Fermi level to the gap leaves relatively few higher-energy states above the Fermi level at the top of the valence band into which the electrons can be selectively shifted (to those states having electron motion primarily in one direction), as pushed by a voltage source, like a battery, for electrical conduction. For semiconductors of type (c), few states at the bottom of the conduction band above the gap are occupied to begin with, so there are few electrons to be pushed by a voltage, and conduction is limited. For semiconductors of type (b), the situation is more complicated.

For semiconductors of type (b), the thermal agitation of the lattice can *excite* a relatively small fraction of the electrons out of the valence band, across the relatively small band gap, and into the *conduction band* of states above the gap where they have higher energies, are not tightly bound, and (*thinking classically now*) are free to roam around the solid. This leaves *holes* of local net positive charge in the valence band since the number of electrons locally surrounding an atom is now less than the number of protons in that atom's nucleus.

A hole can effectively move from one atom to its neighboring atom if an electron of a neighboring atom hops over to fill the hole in the first, thereby creating a hole of its own. The holes can thus wander around from atom to atom, and they can move like positively charged particles under the influence of a voltage that may attract them to a negative terminal and repel them from a positive terminal. (What is really happening is that electrons are attracted to hop toward the positive terminal, leaving the hole to move in the opposite direction.) At the same time, the electron that escaped to roam around in the conduction band can also move toward the positive terminal. So both the electrons in the conduction band and the hopping electrons that move the holes in the valence band move toward the positive terminal. Having negative charges moving *to* the positive terminal is the same as having positive charges flowing *from* the positive terminal, which is the way that current is defined.

Unlike the situation in metals, for this (b) situation the electrical resistance to conduction in these semiconductors goes up as the semiconductor is cooled. That is because the thermal excitation that creates the conduction electrons and holes is reduced or removed as temperature is lowered. It is this property that has been used to make temperature-measuring instruments called *thermisters.*

The stacking of atoms in solids can occur in various patterns. And in silicon and germanium, as in the diamond form of carbon discussed in Chapter 15, the atoms tend to stack themselves in a basic tetrahedral geometry that extends throughout the crystal. The two s-state outer electrons and two p-state outer electrons in the atoms of these elements combine in a hybrid arrangement, extending out to bond with neighboring atoms at what can be visualized as the corners of a tetrahedron. In the diamond form of carbon, the band-gap energy difference between the valence band and the conduction band is much too high for excitation from the one to the other to take place at normal temperatures. And so diamond has no conduction electrons and no holes and is a perfect insulator. But for silicon and germanium, the bonding is not as strong, the valence band is at a higher level, and the band gap is smaller, permitting the semiconductor behavior described above.

DOPING, AND THE CONSTRUCTION OF CHIPS CONTAINING BILLIONS OF TRANSISTORS

Conduction electrons can also be created by substituting some of the silicon or germanium atoms with atoms, like phosphorus, that have five rather than four outer electrons. This *doping* produces a so-called *n-type* material with extra (negatively charged) conduction electrons in the conduction band beyond the band gap. Or these elements can be doped by substituting some of their atoms with the atoms of elements, like gallium, that have three rather than four outer electrons, to produce *p-type* semiconductors that have a starting population of holes (we say positive, because of the electron deficiency), vacant electron states in the valence band below the band gap. Special properties and applications then result when sandwiches of *n*-type and *p*-type semiconductors are put together—as in *diodes*, which let the flow of electrons occur across the boundary between the two types in only one direction, and in *transistors*, three-terminal devices that may have three layers *n-p-n* or *p-n-p* materials in which the control of a small current entering one terminal (layer) regulates the large current flow across the other two terminals (layers), so that electrical signals can be *amplified*.[1] Or information can be stored and retrieved by controlling the flow to be either on or off, representing 0 or 1 binary states, for information storage in billions of tiny, tiny *transistors* in the memory or processor chips of computers.

The integrated circuits of billions of transistors and other elements in these chips are created using a process of photolithography in which dozens of chips are produced at once on up to twelve-inch-diameter wafers of silicon. As many as fifty processing steps are used, repeating a sequence of coating the wafer with a photo-resist that is etched away only in areas exposed to particular ultraviolet light and then treating the regions thus revealed. (The photo-resist is a uniform covering that prevents the deposition of new material or other operations, except where it is exposed to light and etched away, exposing the surface to be treated beneath.) The light is projected or shines through a mask to expose a "blueprint" of minute transistor and circuit geometries on microscopic and nearly sub-optically-microscopic scales, so that the exposed features can be variously treated. In one subsequent step, they may be injected with a particular doping element (*dopant*). In another, after differently masking, exposing, and etching again, another dopant may be added in other regions. Many steps later on, after still differently masking, exposing, and etching again, copper may be evaporated onto the wafer and then selectively removed by coating, exposing, and etching, to make connections between transistors and other circuit elements and electrical "leads" to the edges of the chip for its later connection. Finally, the chips may be cut, sealed, individually encased, and tested for use in computers or other devices.

CHARGE-COUPLED DEVICES (CCDS)

Charge-coupled devices (as nicely described by authors Bruce Rosenblum and Fred Kuttner in *Quantum Enigma*), "have greatly expanded personal photography, revolutionized astronomy and are steadily improving diagnostic medicine. A typical digital camera has a semiconductor chip with millions of CCDs."[2] In something related to the photoelectric effect (identified as such by Einstein) described in Chapter 2 and Figure 2.3, photons excite a cluster of electrons in silicon states, which can then be moved by an electric field giving the location where they were created and converted to measure the intensity of light at that position. As a result of his work, in 2009 Willard S. Boyle was awarded the Nobel Prize in Physics "for the invention of an imaging semiconductor circuit—the CCD sensor."

APPLICATIONS

Most of what we broadly call "electronics" and almost everything that we use in electrical controls and communications involves semiconductors formed to make diodes or transistors that are integrated by the thousands or millions or billions in the chips that are assembled with other types of circuit elements to make devices. The uses for semiconductors and the inventions and products made from them are too numerous to describe here. These include, to name just a few: computers; smartphones; hearing aids; radios; TVs; DVD players; hi-fi systems; modern telephones; printers; scanners; fax machines; the computerized nerve centers in robots, automobiles, airplanes, rocket ships and satellites; GPSs; cell phones; motor controls; lock-in amplifiers and other scientific instrumentation; exercise equipment; radar and sonar; fish finders and depth sounders; automatic pilots; numerically controlled machine-shop equipment; digital cameras; and, with the basic laser element, the multitude of laser applications that are described in Chapter 4, including the barcode reader.

And, of course, we have semiconductor solar panels for the direct conversion of sunlight into electrical power. The utility of this intermittent or highly variable power source may rely on technologies for energy storage through any of a number of means, including one that involves the electrolysis of water to produce hydrogen, as described in Chapter 22 in relation to carbon nanotubes as a storage material.

NEW DEVELOPMENTS

The rate of advancement in the development of semiconductor components and applications is astounding. I'll mention first just a few of the recent developments in chip components.

I have already described in Chapter 22 the development of miniature supercapacitors. Similar developments are reported by Charles Q. Choi in his article "Nitrogen Supercharges Super-Capacitors."[3] In the same News section of *Spectrum*, February 2016, less resistive interconnects within chips is described in "Rise of the Nanowire Transistor" by Richard Stevenson,[4] and the integration of 70 million transistors and 850 optical components into a silicon processor is described in "Linking Chips with Light" by Neil Savage.[5] "Survival in the Battery Business" in *MIT Technology Review* of July/August 2015[6] describes a company built around the development of a solid-state battery. In the same issue, the company highlighted in the article, SAKTI3, was cited among "The 50 Smartest Companies"; along with Imprint Energy, which has developed ultra-thin flexible rechargeable batteries that can be printed cheaply in commonly used industrial screen printers; and along with SolarCity, which is planning to become the Western Hemisphere's largest manufacturer of silicon solar panels. An updated report, "SolarCity's ($750 million) Gigafactory" is reported in the March/April issue of the same magazine.[7]

Month by month, new or improved products are brought to market. The Internet has become a driving force. But in computers there appears to be a classical limit as transistors and other circuit elements begin to approach the size of the atoms of which they are composed. This would seem to signify the end of advancement in the capacity of chips. (See "Transistors Could Stop Shrinking in 2021," by Rachel Courtland.[8])

Meanwhile, the prospect of operating with circuit elements exhibiting quantum behavior has resulted in the early-stage development for some applications of superpowerful *quantum computers*. Projects underway in various parts of the world may revolutionize the industry for those applications, as described in Chapter 8 and Appendix C.

Chapter 24

SUPERCONDUCTORS II—LARGE-SCALE APPLICATIONS IN SCIENCE, POWER GENERATION, AND TRANSMISSION

Superconductors promise powerful and compact electric motors to drive the propellers of ships, and compact generators for airborne defense. For commercial power components—that is, generators, transformers, and transmission lines—superconductors would avoid resistive heating and the associated power losses (inefficiencies) presently found in these devices. Or superconductors could allow us to build more compact components with enhanced capabilities. But the biggest demand for the development of superconductors and superconductor windings over the years has been from funding to develop and build superconducting magnets for huge experimental devices in particle physics. I briefly first mention two such devices and their latest accomplishment.

BIG SCIENCE—PARTICLE ACCELERATORS

Superconducting magnets are essential parts of modern particle accelerators. More than "atom smashers," these huge machines are creating particles mainly out of pure energy. And they employ tens of thousands of tons of superconducting particle-beam-bending magnets, as for example in the building of the Tevatron at Fermilab near Chicago and the Large Hadron Collider (LC) near Geneva, Switzerland, the latter of which is described in some detail in Chapter 9, Part IV (C).

ADVANCED SUPERCONDUCTORS

Large-scale commercial success for power applications depends on our being able to make superconducting devices low enough in cost and sufficiently reliable that they will pay for themselves on the long term. That means that the cost of refrigeration to keep superconductors operating below their critical temperatures must be less than the dollars that are saved by replacing conventional devices to eliminate the inherent resistive heating in their copper windings and their associated power losses. And superconductors are absolutely essential if the *tokomak* fusion reactors under development are to generate more power than is consumed by the magnets that confine the fusion plasma, as described in Chapter 20.

Superconductor operation at higher temperatures tends to improve overall economic performance because refrigeration equipment is much smaller, less complicated, less expensive and more reliable for higher temperatures and because refrigeration costs increase in more than inverse proportion to the absolute temperature at which the superconductors operate. So, for example, refrigeration for superconductors that can operate in liquid nitrogen[1] at its boiling point of 77 Kelvin degrees (above the absolute zero of temperature) will cost less than one twentieth as much as refrigeration for superconductors that operate in liquid helium at about 4 Kelvin degrees. (Note for reference that a normal room temperature of 72 Fahrenheit degrees = 22 Celsius degrees = 295 Kelvin degrees above the absolute zero of temperature. So, liquid nitrogen boils at a temperature a bit more than a quarter of the way from the absolute zero of temperature toward a normal room temperature, and liquid helium boils at a temperature just a little less than 1/50 of the way to room temperature.)

Refrigeration becomes particularly important for AC power generation and delivery applications. This is because superconductors, though they have no resistance if kept below their critical temperatures, do generate some *hysteresis* and *eddy-current* losses (i.e., generate heat) when operated in changing magnetic fields. Superconductors used in the generators, transmission lines, and transformers operate in the presence of changing magnetic fields. So the savings that might be made by using superconductors to get rid of resistive losses in these devices must be weighed against the cost of refrigeration to expensively remove the relatively small amount of heat that is still generated.

New, so-called *high-temperature* superconducting materials (HTS materials), and even *second-generation* HTS materials, have been developed that can be used in relatively inexpensive liquid nitrogen.

The most commonly used so-called *first-generation* HTS material is $Bi_2Sr_2Ca_2Cu_3O_{10+x}$ (bismuth-strontium-calcium-copper oxide, BSCCO for short). (Some chemistry!) While BSCCO can operate at liquid-nitrogen temperature, the magnetic field levels to which it can operate and the amount of current that it can carry in a superconducting manner at liquid-nitrogen temperature is limited. This is particularly so as compared to second-generation HTS materials, typically involving $YBa_2Cu_3O_{7-x}$ (yttrium-barium-copper oxide, YBCO for short) or $REBa_2Cu_3O_{7-x}$ (rare-earth-barium-copper oxide, REBCO for short). At the moment, both of these HTS materials are still relatively expensive as compared, for example, to the NbTi alloy superconductor currently used (in liquid helium) for MRI.

A compromise between low- and high-temperature operation is offered by an intermediate temperature ($Tc = 39$K) MgB2 superconductor that has been developed in recent years. One review also cites particular applications for which this superconductor is suited.[2]

POWER APPLICATIONS, OF A SCALE TO LIGHT CITIES

Efforts to develop superconducting components for commercial electric power have been pursued worldwide for over fifty years. But to my knowledge none of these devices have as yet been accepted for widespread commercial use. Some development work still continues, but it is much diminished compared to what was being pursued even fifteen years ago.

The key to success will be to make superconducting devices low enough in cost and sufficiently reliable that they pay for themselves on the long term. Just one day of unscheduled "down time" can cut drastically into the dollars that would be saved through the improved efficiency and the operational advantages of these more complex superconducting devices. Competitive cost and reliability will be essential. Even if short-term tests are completely successful and show superconducting devices to be economically and functionally very attractive, it may take many years to demonstrate the reliability demanded for acceptance by an appropriately conservative power industry.

To give you some sense of the effort, I describe below some typical projects, several of which I either led or worked on at various stages in my career: as a materials scientist, physicist engineer, or project manager.

Fusion Reactors

The absolute need for superconductors in magnetic confinement types of fusion reactors is described in Chapter 20.

Superconducting Generators

Although eventually superconducting generators were constructed to power levels of 70 MVA (for our purposes 70 megawatts, or 70 million watts), I cite here an earlier development in which I participated, as both designer of the superconductor and foreman and designer of the instrumentation that was used for the test.

Figure 24.1 shows the 5 MVA superconducting generator that we built and successfully tested at Westinghouse in the early 1970s.[3] At the time, many of us thought that superconductors would revolutionize the power industry. The 5 MVA was the most powerful superconducting generator built until that time. (The generator is located behind the engineers standing to the center and left. I am the bearded fellow who is standing to the far right.) I show this figure to give you some sense of the compactness of a machine that is capable of providing the power to light five thousand homes.[4]

Fig. 24.1. A 5 MVA (for our purposes, 5 megawatt) superconducting generator after successful test in 1972, with some of the Westinghouse design and test team alongside. *Left* to *right*, in the front row, are Jim Parker, Don Litz, Adolphus Patterson, Cliff Jones, and Tom Fagan; standing, are John Mole, Henry Haller, and author Mike Walker. (Image reprinted with permission from R. D. Blaugher et al., "Superconductivity at Westinghouse," *Superconductivity News Forum* 6, no. 20 [2012]: 12.)

Project Organization and Partial Department of Energy Support

As the development of large-scale power components advanced sufficiently, demonstration projects were started (and some have been completed) to place superconducting power components into actual use in commercial operating situations. In the United States, these demonstrations have often been accomplished in team efforts involving a superconductor manufacturer, a device manufacturer, universities with specific relevant expertise, and an electric-power company that would put the hardware into use, along with engineers and scientists from one of the Department of Energy's major laboratories. DOE partially supported and monitored many of these projects, and the projects normally underwent an annual peer review.

To give you some sense of the scale and purpose of these projects, I mention two of them, those familiar to me mainly from my involvement as engineer and project manager at IGC's SuperPower division, near Albany, New York.

Transmission-Line Cables

This cable was constructed, installed, demonstrated, and then retired. (I had only a preliminary involvement in the project's getting started, just before I retired in 2002.) It was a nearly quarter-mile-long *HTS underground transmission-line cable,* installed and operated for many years in a standard utility right-of-way connecting two National Grid substations in Albany, New York. One impetus for superconducting cable development is the present 7 percent to 10 percent loss of electrical power caused by the electrical resistance of conventional copper transmission-line cables. A further advantage is that each superconducting cable can carry three to five times the power of a conventional cable in the same underground duct space, so that the substitution of HTS cables into present cable tunnels may provide increased power without the need for increased excavation and rights of way. (This is a really big deal in the fight for crowded underground space below some major cities!)

Our project's flexible, vacuum-jacketed superconducting cable was installed and operated initially in 2005 using a first-generation BSCCO HTS superconductor. In 2007, thirty meters of the cable was replaced with a higher-*Tc* YBCO second-generation HTS superconductor, also demonstrating the ability to make joints and successfully operate with these joints. Both sections use liquid nitrogen as a coolant. The line carries 800 amperes in each of its three phases, each at *34,500 volts* for a power delivery of 48 MVA (that would be *48 megawatts* if it were all resistive power, enough to light 48,000 average homes).

Partners with SuperPower on this project were Sumitomo Electric Industries, Linde, and National Grid. Funding was supplied in part from the New York State Energy Research and Development Authority (NYSERDA) and the US Department of Energy.

(I note a recent such development: A fault-current-limiting superconducting cable has been bringing power into City Center in Essen, Germany, on the grid since March 2014. It delivers up to 40 MVA (for our purposes, think of 40 megawatts) at 20,000 volts over a distance of one kilometer [0.6 miles].[5])

Fault-Current-Limiting Transformer

Another project, one that I started and led through most of its first (transformer only) phases, has morphed and advanced considerably into a program to build a combined *HTS superconducting-fault-current-limiting (SFCL) transformer*. This superconducting transformer promises to be smaller, lighter, quieter, and more efficient than conventional transformers. It should be able to operate during faults[6] at above-rated currents without danger to transformer life. With liquid nitrogen as an electrical insulator and coolant, rather than conventional transformer oil, the HTS transformer will not be a potential fire hazard. (After all, nitrogen does not burn, and it already comprises 80 percent of the air, so the release of nitrogen into the atmosphere is also not a problem.) And with the added fault-limiting capability, this transformer would provide downstream protection for substation circuit breakers and greater upstream grid flexibility, all without any negative impact on overall grid performance.

The objective of this project was to design, develop, manufacture, and install on a live utility host site a medium-power smart-grid-compatible superconducting fault-current-limiting transformer. This transformer would step down[7] an incoming *69,000 volts* to an outgoing *12,470 volts* while delivering 28 MVA of power (that would be 28 megawatts if it were all resistive power). The SFCL transformer will use second-generation HTS superconductors cooled and insulated in liquid nitrogen. The project was slated for completion in 2015. It was funded in part by the US Department of Energy. Partners with SuperPower on this project were Waukesha Electric Systems (formerly Waukesha Transformer), Southern California Edison (SCE), Oak Ridge National Laboratory, and the University of Houston (TcSUH). (Rochester Gas and Electric and RPI, Rensselaer Polytechnic Institute, were partners through the earlier phases before SCE and TcSUH came on board.) Waukesha experienced a change in management and chose to withdraw from the project, so it has continued only as more of a proof of concept than a transformer demonstration.

Here in the United States, approximately 140,000 medium-power transformers are approaching forty years of service, that is, they are nearing the end of their useful life. They will soon need to be replaced, and the development of this new and improved superconducting transformer technology, if it is successful, would be timely for an updating of this component throughout the entire US power grid.

These are only a few of the projects being pursued by just a few companies developing materials or applications for superconductors. By now you have a sense of the difficulty in achieving success in the power sector. Though important contributions are being made, we need to soberly evaluate the status and possible overall impact of these developments, and we must recognize that acceptance by a necessarily conservative power industry does not usually come quickly. But much is being done in superconnectivity overall. To give you an idea of work being done by industry, governments, and universities worldwide, note that the Applied Superconductivity Conference, which is held every two years, is just one of several major conferences at which new or ongoing work on superconducting materials or devices is reported. Typically well over five hundred presentations are made to report works in progress at each ASC conference.[8]

Remember, we have considered in Part Five only some of the many modern inventions under development or in operation that either have been spurred by or are properly understood through our quantum view of the world around us. What we have learned and produced in the last seventy years, aided by our understanding of quantum mechanics, is truly amazing. What we do in the next twenty years will boggle the mind!

ACKNOWLEDGMENTS

This book has been shaped substantially by the suggestions, questions, and comments of many reviewers and advisors. I am grateful to all for their help. Ultimately, however, I made the choices as to content and I am responsible for any errors of fact or interpretation that may have occurred. I sincerely apologize to anyone who may have been inadvertently omitted in these acknowledgements.

First, I thank Mr. Steven L. Mitchell, editor in chief, and his excellent staff at Prometheus Books, for suggesting a broader scope, for encouragement, and for patient guidance. Producing *Quantum Fuzz* has been a team effort, starting with Catherine Roberts-Abel's flexible accommodation of both my schedule and this book's special needs in layout, then Hanna Etu's diligent and effective help in soliciting both permissions and reviews, Jacqueline Nasso Cooke's obvious artistry in design of the cover, Jade Zora Scibilia's understanding of the book in her thorough but reasonable interchanges through copyedit, with associate editor Sheila Stewart, indexer Laura Shelley, and typesetter Bruce Carle, and finally Jake Bonar's quick responses in educating this author to the needs of marketing and publicity.

The book has benefitted greatly from the extensive review and advice of Dr. David Toback, Thaman Professor for Undergraduate Teaching Excellence and professor in the Mitchell Institute for Fundamental Physics and Astronomy at Texas A&M University. Dave is author of the recently published *Big Bang, Black Holes, No Math*. His enthusiastic response to *Quantum Fuzz* was especially encouraging, and he has graciously written its foreword. L. Howard Holley, fellow physicist, recently active in the key biology-related study of protein folding, not only reviewed the physics but also added particular depth and background to the discussion of the bonding and shape of the water molecule. Longtime friend Delia Milliron, associate professor of chemical engineering at the University of Texas–Austin, looked over Part Four, dealing with the quantum mechanics of the atom and the foundations of chemistry. John Gardon, vice president of research first at Akzo-Nobel and then at Sherwin-Williams, now retired, kept me out of trouble when I ventured more deeply than I should have in that direction. John has been a continuing source of information on the development of energy resources and technology.

The book owes its title to Fred Bortz, friend since graduate school, punster extraordinaire, mentor, and successful author of science and technology books for young readers. Fred provided critique of both the physics and the writing. "Dr. Fred," as he is known, is also a reviewer of science and technology books for major newspapers. Dr. Leslie Polgar, another friend from graduate-school days, physicist, business executive, and now adjunct professor and consultant looked over Chapter 8 on quantum computers. And John Gribbin, author of *Computing with Quantum Cats* and, just recently, *13.8: The Quest to Find the True Age of the Universe and the Theory of Everything*, reviewed and advised on the first half of the book. My cousin C. Terrence Walker provided early reviews, stimulating discussion, quotes, and recommendations of related reading on philosophy and the nature of quantum mechanics. Terry has a PhD in intellectual history and a special interest in the philosophy of assumptions in science.

Barbara Oakley, translator on Soviet trawlers, South Pole radio operator, army captain, doctor of systems engineering, professor, and bestselling author (*Evil Genes—Why Rome Fell, Hitler Rose, Enron Failed, and My Sister Stole My Mother's Boyfriend*) taught me a bit about writing, offered encouragement, and provided strategic guidance as I moved toward publication. It was Barbara who recommended Will DeRooy, excellent independent copy editor of an early draft, whose advice and questions and comments regarding structure and meaning went well beyond his assigned task. Jack Kaufman aided with skilled photography, and Martha Thierry provided timely, careful, and quality work on all of the illustrations that I generated. Kathy Braun, a professional at co-writing and coaching prospective authors, asked questions and made suggestions. Professor Joe Grimm of Michigan State University taught me much and offered advice and encouragement in his local short course on self-publishing.

My colleague and friend from Westinghouse days and since, Alex Braginski, provided me with an update of developments in superconductivity just before editing. And Art Wiggins, emeritus professor of physics and author of *The Human Side of Science* both reviewed and advised at this final stage.

One major objective in my writing this book has been to bring to the general public a sense of the inherent beauty in the structure of the atom, the building block of all that we see around us. The challenge has been to simplify and present in a nonmathematical way the quantum mechanics that gives us this view: a theory that is counterintuitive and inherently mathematical and abstract. To meet the challenge, and to be sure that I am describing our quantum world generally in an interesting and readable way, I enlisted the help of many friends and family members as the book grew over the period of a dozen years, constructing draft after draft, each submitted in whole or in part to some few for review and critique. Here too, I am grateful for all of the help that I have received.

My initial critics were my immediate family, starting with my late wife, Roslyn: partner for thirty-five years, social worker, and perhaps the most commonsense wise person I have ever known. (Though she was generally supportive, her initial assessment was two thumbs down!) My sister, Johanna Darden, has been encouraging all along. (She is perhaps atypical, with more math and physics than most of my prospective readers.) Her husband, Bill (once my roommate at MIT and subsequently professor of linguistics and Slavic languages at the University of Chicago) provided a more philosophically oriented discussion of quantum theory. My son Doug, a graduate in mechanical engineering from MIT and then the University of Michigan (and my daughter-in-law, Liz, in computer science from Wellesley) read through and critiqued several early and parts of later drafts. Doug noted that he had gained a better feel for what he had already been taught.

I owe a very special thanks to Barbara Halpern. Barbara and I have been together for most of the eleven years since Roz died. She has read through, marked-up, and critiqued virtually every draft produced in that time, and she has continually encouraged the work (though her enthusiasm may be borne in part of an unusual thirst for knowledge and delight in learning). I also thank Eitan and Ariel, Barb's grandsons, for finding the time between school, practice on the cello, video games, and other teenage activities, to give parts of more recent drafts a serious read and commentary.

I benefited greatly from the questions and comments of one of my late son, Arlan's, best friends, Brendan Boyle, former high-school teacher and now computer guru, who offered encouragement and advice after reading through and reviewing each of several drafts. I am also appreciative of reviews by Bobbi Markiewicz; Ed Coe; Chuck Townsend; Bill Ahlstrom; Bernie Stuecheli; Joe Reynoso; Frank Mandelbaum; Emeritus Professor Donald Shuster; and my cousins Geraldine Spilman, Marc Landgarten, S. Marc Tapper, Stanton and Jai Walker, Richard and Carole Walker, Leslie and Lorne Kermath· Jacob Hodges, and Steve Horelick, who either marked up drafts or sections substantially, wrote separate comments, asked key questions, or provided verbal comment based on a substantial read. It was Steve who triggered my description of electron states in the atom in analogy to the standing-wave patterns in musical instruments.

Others who provided encouragement or discussion or looked at drafts or parts of drafts and gave comment and advice include: Mark Silverman; John Ricci; Bob Brewbaker; Tom Matz; Beverly and Jerry Viedra; Marlene Karp; Fred Garon; Irene Stein; Bill Margolin; Judy Kaufman; Priscilla and Bob Pettengill; Bob Agel; Al and Martha Devernoe; Ana Borzha; George and Karen Schnakenberg; Burt Halpern and Anna Zayachkowsky; Barbara's brother, Larry Jeris, his wife, Karen, son, Justin, and son-in-law Spiro Xydas; and my Walker cousins Jim and Harriet, and their daughter Ellen and her husband, Neil.

Finally, I thank Margaret L. Leighton for permission to use key figures from her late husband's book *Principles of Modern Physics* (Reference F), Dr. Leonard W. Fine for figures from his *Chemistry for Engineers and Scientists* (Reference E), and Sidney Harris for the five cartoons that you find strategically located to provide a lighter touch.